Packaging and Ecology

Packaging and Ecology

Packaging and Ecology

Frans Lox

Pira International

ISBN 0 902799 93 2 (hardback)
ISBN 1 85802 013 1 (student edition)

Published by
Pira International
Randalls Road
Leatherhead
Surrey
KT22 7RU
UK

Tel: (+44) 0372 376161
Fax: (+44) 0372 377526

This book is a revised and updated version of *Verantwoorde Verpakking.* © Copyright Stichting Leefmilieu vzw, Keizerstraat 8, 2000 Antwerpen, 1983.

Typeset in the UK by Frost Design Associates,
West Drayton, Middlesex.
Printed in the UK by Antony Rowe Ltd,
Chippenham, Wiltshire.

'Consumers, by definition, include us all'.

'Just as consumers have the right to know what is in their credit contract, so also do they have the right to know what is (in) the package they buy'.

John F Kennedy, 15 March 1962

This book is dedicated to:

- Professor Frank Paine, secretary general of IAPRI, for his friendship and most appreciated guidance into the field of packaging,
- I.K. Ilysom for providing help in discussing the problems,
- my grandchildren, who hopefully will appreciate packaging and learn from its advance.

Contents

Foreword

While efficient packaging is a necessity for every kind of product and is an essential link between the producer and the customer, the emphasis on several of its functions has changed in the last 20 years.

The basic function of packaging is to identify the product and ensure that it travels safely through the distribution system to the final user. Packaging designed and constructed solely for this purpose adds little or nothing to the value of the product: cost-effectiveness is the sole criterion for success. If, however, the packaging facilitates the use of the product, is reusable or has an after-use, some extra value can be added to justify extra cost and promote sales. Many examples of packaging providing such extra value can be cited.

Obviously, the detailed packaging requirements will be dependent on the product and the distribution system, but there are certain points which must always be considered; in almost every instance it is the product alone that the customer wants – the packaging, with a few notable exceptions, is an ancillary part of the transaction. And while customers recognize that the packaging must be adequate, an appearance of over-elaboration may well create the impression that the cost of the product has been unnecessarily raised by the packaging; this aspect can result in a negative effect on sales. Consumers' requirements can be summarized as follows: the packaging should be easy and safe to handle, simple to open and use, and at least provide few problems for the environment.

Thus, packaging is a constituent part of the socio-economic system and provides an important link between industry, trade and the consumer as well as facilitating distribution. The vast majority of people recognize packaging as a social good in respect of its protective functions but many still have doubts about those related to the ecology and a few regard it as socially undesirable.

In this book, Dr Lox examines the whole range of packaging interactions from technical and social standpoints and discusses the new relationships between man, machine, material and the environment, pointing out that in the future only socio-ecologically acceptable packaging may be tolerated.

Frank Paine

Preface and acknowledgements

For centuries, packaging has been an element indispensable for the handling and preservation of goods. Nowadays, almost every product is supplied pre-packaged, especially in the retail sector. Moreover, each item of packaging has been thoroughly studied and tested for the purpose it has to fulfil during transport, storage and distribution; there is no doubt it can give a high-tech look.

The outline has also been carefully designed because form and colour have gained an important impact on the product's saleability: these visual aspects, together with printed information, have become the most direct means of communication between the manufacturer and the consumer. Although the advertising is subject to rules for the sake of the consumer, the meaning of abbreviations, marks and symbols remain for most unintelligible. And while packaging regularly reaches the news headlines, most consumers may have the impression they are faced with an 'obligatory packaging', adding 'unnecessary costs' to the product and producing 'ecological problems'.

The introduction of new materials and the continuing tendency for non-returnable packaging, plus the high prices of raw materials and energy, have produced new problems in that measures to recover materials and energy from packaging refuse and to keep it out of the household waste bin, have been proposed on a European scale and/or applied on a local scale.

Hence this book is designed to highlight packaging, its economics and raison d'être, to provide a survey showing the complexity of its functions and the interactivity of the good's requirements and the properties of materials, to examine energy and transport costs, the safety regulations and the measures for waste management. The relationship between environment, the manufacturing process and the disposal of packaging is also emphasized.

As far as packaging is concerned, every consumer can make a sensible and responsible choice: all relevant themes are discussed and summarized under the chapter 'Justification for packaging'.

The opinions on packaging presented in this book are mainly derived from theoretical and practical experience: the initial concept is a Flemish-written monograph *Verantwoorde Verpakking* (*Packaging Justified*), number 22 in the series edited by the Ecological Foundation of the Kredietbank-Belgium. Tables have been updated and new ones concerning eco-profile and life-cycle analysis have been added. The original figures have also been updated where appropriate. The whole text has been revised and updated with sections on 'child-resistant and tamper-evident closures', principles of eco-balances, life-cycle analysis and eco-profile, with a list of European Directives concerning packaging safety and the policy of packaging waste. Despite these changes, the objective of the book is still to examine and to explain all aspects of today's packaging industry, and to give a critical analysis of the actual problems which arise.

I wish to express my gratitude to all who have contributed – directly and indirectly – in the realization of this book: to the members of SEAL (Synergiae of European Analysis Laboratories) stimulated by their president Professor Dr M. Balzon for the decision to mandate this English version and to Mrs Ir Chr. Nottebaert for the carefully provided translation, the framework of this book, my deeply sincere feelings of respect.

To Professor Frank Paine, my grateful thanks for his spontaneous offer to review the whole new version and for his most valuable suggestions, comments and help, especially in establishing the glossary with clear definitions on packaging and ecological terms.

Many people have unselfishly contributed in the achievement of this work: to Ph. I. Milewski for the helpful discussions and the layout of the chapters, to Ir J. Boone, M. Buelens

(Tetrapak-Belgium), Drs E. Nierynck (Free University Brussels), Ir Scheys (Dir. at Fechiplast), V. Machiels (Dir. at B.P.I.) Ir L. Debleeckere (Eurodeksels) for providing the data and information on packagings, to all personnel of both the Institutes of Science and Technology (University of Gent) and Belgian Packaging Institute, Brussels for their willingness, my sincere and well-meant thanks.

My sincere appreciation to my daughter Dr H. Lox for her devotion in managing the software of the text, and my thanks to my wife and members of the family for coping with my absences. Last but not least I would like to thank all members of the editorial board at Pira International for their efforts in the creating of this book.

Frans Lox

I
Packaging

1 DEFINITIONS

Packaging includes all those physical elements used in wrapping goods and assisting their handling. The need is to maintain by an appropriate packaging the initial product quality – obtained at the end of the production process – during the further preservation, storage, and transport operations. However, the packaging process itself is not a method to increase the natural life of the product, but a means of maintaining that quality. Packaging always separates the product from its surroundings; this separation raises a barrier and creates an 'internal' environment, which should avoid any biological, physical, mechanical and chemical transformations, exchanges and/or potential interactions that could deteriorate the product (see Figure 1.1):

- exchanges between the internal environment and external surroundings,
- exchanges between the packaged product and the packaging material (migration phenomena).

The psychological impacts due to shape, dimension, colour and printed information are emotional and intangible elements included in packaging. The saleability of a product depends on these, while the packaging material components are functional and indispensable. Marks and symbols can be printed on the packaging for automatic computer-aided data processing. Packaging is not limited to the external visible container but also includes internal components designed for quality preservation. For example, three layers of lemons placed in a wooden

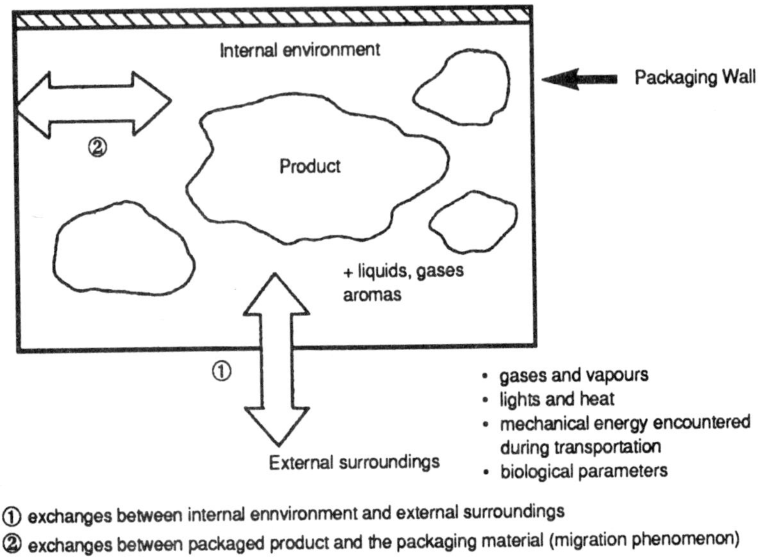

Figure 1.1 Packaging serving as a barrier between the internal environment and its external surroundings

case are better protected against mechanical shocks during transport if a tray insert (e.g. of foamed polystyrene) is placed upon the bottom; each layer then lies well ordered (following a centred cubic model), resulting in an optimal distribution of the compression stresses on the contact points (eight for the middle layer). To prevent damage during shipment, electronic equipment can be fixed on a wooden fitting by means of rubber shock absorbers and afterwards slipped into an impermeable bag containing a desiccant such as silica gel. Finally, to facilitate stacking, the assembly can be placed in a wooden case or in a strong fibreboard box. Ropes, straps, caps, inserts, inks, labels etc. can be considered as packaging accessories, indispensable for the total packaging assembly. Even gases can serve as packaging accessories when they are specifically used to preserve the quality: e.g. oxygen for packaging raw meat, or nitrogen to obtain an oxygen-free and inert atmosphere. On the other hand, each propellant gas, as used in aerosols, is part of the packaged product itself, as is the carbon dioxide in sparkling water and beer. In these last cases, gases are not packaging accessories, but product components.

Each unit package contains a predetermined or stipulated amount of goods expressed either by number, weight (in kg) or volume (in litres). This quantity will preferably be adapted to the average consumption rate during the period allowable for quality preservation. Consumer packaging usually consists of single units, but may sometimes be multi-unit packages when a number of primaries are collated together to promote more efficient stacking and less handling during transport and warehousing. Such collective packaging is useful in the case of periodic purchasing (weekly, monthly). So a packaging containing 12 cartons of UHT milk is a collective multi-packaging as well as a consumer packaging; both the modular dimensioning, and the longer keeping period of each packaging unit (three to six months) are important.

The term 'transport packaging' refers to the additional packaging that is given to a number of unit (primary) packages in order to reinforce their protection during transport by land, sea or air. Sometimes primary units are also designed to be transport packaging (cf. packaging of electronic devices, TV sets or cameras).

Packaging involves three main operations (see Figure 1.2). The operations are related to the realization of:

1. packaging manufacture (e.g. conversion of paperboard sheets into boxes; blowing glass bottles from molten glass, etc.),
2. filling,
3. closing and/or sealing.

Thus packaging items are not just containers such as cups, bowls, plates, pans, etc. because they are closed or sealed to give a guarantee to the consumer and the manufacturer. Even if this seal has been broken to open the packaging, a certain tightness can often be restored (e.g. stoppers on threaded bottles and jars), but the full sealing can never be recovered.

Once opened, a package loses saleability, the wear and tear period starts and the guarantee on the product is limited. Packaging in this connection can be considered as a one-way package, a container able to meet its entire packaging function once only.

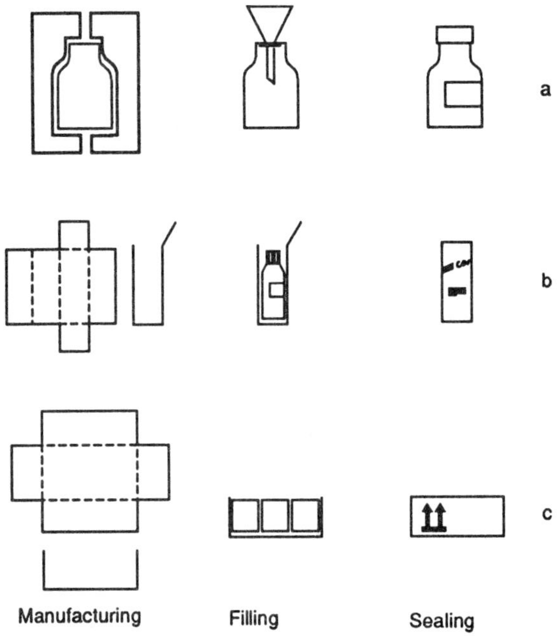

Figure 1.2 Diagram showing the three basic operations in goods packaging: manufacturing, filling and sealing: (a) a unit package; (b) an assembled unit package; (c) a collective package, potential transport package

Figure 1.3 is a diagram of the packaging concepts shown in sequence. Packaging becomes multi-trip (returnable) if it is possible to provide after each use the same quality level as for new packaging. Glass bottles can be reused several times (the average for milk bottles is 15 to 30 times) provided they can be well sealed (crown corks for drinks, bark corks covered with foils for wines) after washing and refilling. The decision to use either multi-trip or one-way packaging depends mainly on the

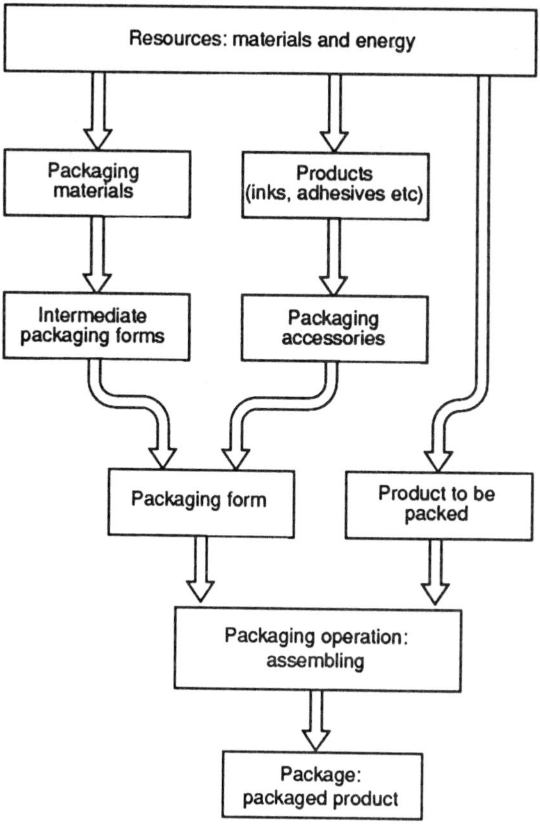

Figure 1.3 Sequence of packaging concepts

cost of managing their reuse. So, in the food sector, it is less expensive to use new glass jars taking into account the total cost of reusing them. Recycling is quite different from reuse of packaging because in recycling the materials must go through a process producing new packaging. Due to the way it is made, an empty food can will never be reusable, it must always be recycled (see Figures 1.4 and 2.2). Throwaway packages are one-way packages, such as food cans and plastic or paper bags, where reuse cannot be considered. The difference between throwaway and one-way packaging is often tenuous and questionable: generally speaking, all throwaway packages are one-way but the opposite is not always valid.

Because of their construction, bulk packaging such as containers and tanks, are generally returnable packaging. Moreover, quantity and quality are guaranteed as they can be sealed (e.g. by customs in the case of international transport). Theoretically speaking, they are of the same kind of material as consumer packaging, however, they are not designed for extensive use and thus do not create environmental problems. Therefore we will not debate them separately. Figure 1.4 outlines the importance of packaging in the transport chain.

Finally, transport such as trucks, lorries, goods wagons, ships and aeroplanes which also serve as containers, are not to be regarded as packaging because their main function is the movement of goods and not goods protection. It is evident that modern packaging is complex and its evolution must be parallel to the modifications of the goods and of transport and distribution systems. Today, packaging has a sociological dimension, as a product closely related to the level of the development and life style of the society it reflects.

2 PACKAGING EVOLUTION AND ECONOMIC IMPORTANCE

a) Packaging evolution

The first packaging materials used by men were probably leaves, skins, clay, grasses and round wooden pieces used for weaving baskets, etc. They were used to wrap goods (protective function) and to make baskets, bags, bowls and pots (protective and transport functions).

The evolution of civilization is based on experimentation, and the development of packaging is no exception. It is chiselled by the evolution of humanity, the origin of which dates far beyond written history.

Among the oldest packaging forms are those of natural origin such as coconut shells, the dried peels of fruits, e.g.

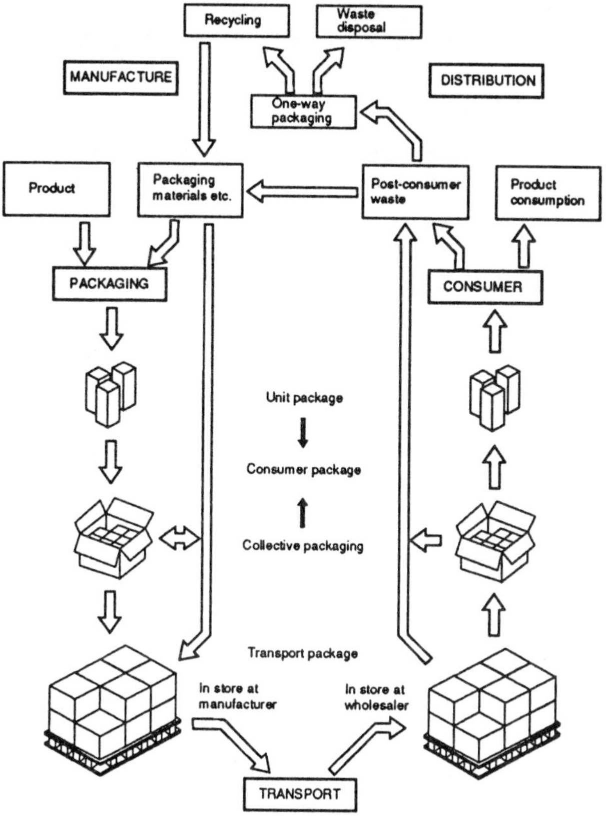

Figure 1.4 Life-cycle of single trip and returnable packaging through production, distribution and retail/consumer sectors

gourds, and hollowed cavities in tree trunks and rocks. These forms could perhaps be sufficient during the Paleolithic age ($1.8.10^6$ BC until the Ice Age) when people ate where food was found. With the exception of copying the cavities carved in wood and rock by natural erosion, the very first manufactured products were produced by converting furs, hairs and skins. Felt, for example, was first made out of animal hairs; much later we discovered synthetic yarns and converted them into cloth. Similarly, twigs and grasses were woven to obtain mats and baskets. We believe the first closed baskets were woven in Ancient Babylonia. Excavations have revealed that during the Mesolithic age (the Ice Age until 8000 BC) people used such baskets to preserve excess food.

While domesticating fire, men discovered the art of pottery during the Neolithic age (between 8000 BC and 4000 BC); later on came the processing of metals such as copper, bronze and iron. Besides their utilization in jewellery (mainly in metals), these new materials were used to manufacture various packaging forms. For reasons of simplicity and accessibility, earthenware remained the main packaging material for quite a long period. The shapes of the well-known jugs and amphorae of Ancient Greece might have been inspired by objects made of wood or woven materials created by the ancient eastern populations in the mountain valleys during 7000 BC (the centre of this culture was the town of Chatal Hüjük in Turkey).

Herodotus (480–430) reports in his *Historiae* Book 2 that, in 454 BC, amphorae were used by the Persians to supply the defeated Egyptian towns with water and wine (cf. Persian wars 498–449 BC). After use, they were even picked up and shipped back to be reused. Based on searches in the Egyptian pyramids, it seems the merchants of that time (circa 2700 BC) used not only pottery forms but also wooden cases, linen sheets (linen yarn) and papyrus leaves for the carriage of goods. Expensive cosmetics and perfumes were kept in little jugs, made mostly in alabaster. The Sumerians (circa 3000 BC) were already familiar with glass; small glass bottles dating from the 18th Egyptian dynasty (1580–1314 BC) show great skill, experience and knowledge of the material. The Romans and Greeks also used glass bottles and stoneware jugs; although, as bottles and jugs break easily, they preferred leather bags for packaging larger quantity of liquids (water, wine, oil) or bulk products (grains). In the Bible, Christ the Nazarene mentions leather as a packaging material: 'Thou shalst not pour young wine into an old skin container as it will blow out and the wine get lost' (Matthew IX,17).

According to Pliny the Elder (23–79 AD) in his *Naturalis Historica*, the wooden casks invented by tribes living in the Alps consisted of staves assembled with iron hoops. At that time, the utilization of iron nails to secure staves to make casks was limited because of their high price: therefore wooden dowels were used. The state of packaging during the feudal Middle Ages is given in Table 1.1: in the slow development since their origin, we notice again the parallels between types of packaging and the civilizations which created them (146,147)*.

The will of people to increase mobility was certainly an incentive to develop appropriate mobile packaging. Since the 18th century with the discovery of sea routes and the extension of trading, wood, glass and metal were used more and more as packaging materials.

Table 1.1 Summary of ancient packaging materials and wares (146)

Material	Packaging form and use
Leather	Wrapping, bags, sacks and bottles
Cloth	Wrapping, bags and sacks
Wood	Boxes, casks and kegs
Grasses and split round wood	Baskets and mats
Stoneware	Small pots and jars
Earthenware	Pots, pitchers, jugs, urns and bowls
Metals	Containers, bowls and pitchers
Glass	Amphorae, jars, pitchers and beakers

Similarly, the technological progress and the increased mechanization of industrial processes during the second half of the 19th century had an impact on packaging: paper, paperboard, tissue paper (unsized paper), glass and metals (tin plate) came into use instead of ceramic wares.

As already mentioned, for the last 50 years plastics have been adopted more and more as new materials for packaging applications. One of the main factors responsible for the modifications in the utilization and structure of packaging forms, is the low production cost of packaging materials and

* The number in parenthesis refers to the list of references (page 279)

forms when produced on an industrial scale; so, paper and board became important packaging materials after the invention of the paper machine to process cheap raw materials. The boom in plastics similarly dates from the period when the need for high-quality petrol (high-octane petrols for highly efficient motors with an increased compression ratio) appeared. The cracking of crude oil not only produced the desired high-quality naphtha but also cracking gases which were first burnt as waste. 'Oil refuse' became a marketable and cheap raw material for the production of a wide range of polymers (e.g. polyethylene: see page 127). Besides the outstanding properties of this new range of products, the space-saving effect of synthetic materials had an influence on their extensive application: plastics for packaging (except films) are generally handled and stored in bead form; the container (bottle, pot, jar) is locally moulded and filled at the end of a production line (e.g. packaging of milk, cooking and lubricating oils) – there is no need to stack 'empty space' in warehouses as is the case with glass bottles and tin cans.

The most complicated and sophisticated packaging shapes, printing and/or components packagers could possibly dream of are now feasible. As far as material properties are concerned, the working capacities of high-speed automatic packaging machines are almost unlimited, consequently more attention can be given to the rationalization of packaging and to the development of new practical functions. The main groups of packaging forms available on the market nowadays are listed in Table 1.2.

Table 1.2 shows that almost any material can be used for packaging: with the exceptions of glass, ceramic and wooden packagings, most of them are made out of 'sheets' of paper, paperboard, metal or plastic. There is normally no specific relation between material and shape.

Packaging can also be grouped according to the package content (Table 1.3). For each, the varying requirements the package must meet are related to preservation for food, to safety for pharmaceuticals (e.g. bottles with a special opening; ampoules), to fragility and damage for appliances, to appear-

Table 1.2 Main groups of packaging forms

1	Boxes and trays (carton, corrugated fibreboard, wood, tinplate, plastic)
2	Wrappers, foils and films (paper, metal, plastic)
3	Bottles and jars (glass, ceramic, plastic)
4	Ampoules (glass)
5	Tubs (glass, paper, plastic)
6	Sleeves and pipes (metals, paperboard, plastic)
7	Containers (metal, plastic)
8	Tubes (metal, plastic)
9	Thermoforms (e.g. blisterpack) (plastic)
10	Bags and sacks (paper, plastic, natural and synthetic yarn)
11	Cases, drums, casks (wood, metal, plastic)
12	Space filling (paper, wood wool, cotton wool, plastic)
13	Bowls (paper, paperboard, plastic)

ance for cosmetics, to nuisances for dangerous goods, etc. Quality preservation is a requirement common to all. Thus the choice of packaging material for a particular application is based on requirements determined by the product to be packaged, by the shape, by the transport means and by the price.

Well-packaged goods are sometimes identified with long-life products designed for long conservation. In the food sector, these aspects are indeed necessary. Progress in the fight against food shortages due to rotting has been made. The Frenchman Nicholas Appert discovered in 1809 that food could be preserved by heating it in a sealed container, and in 1810 metal cans were made by the Englishman Durand. Later, Pasteur (1822–1895) discovered that this preservation was due to the destruction of microbes by the heating process.

For many marketable goods, a knowledge of their characteristics and the factors altering their quality (influence of light, atmospheric oxygen, water, etc.) is the first requirement in the design of a package that will ensure a reliable preservation of quality during transport and warehousing. In this connection and especially in food supply, packaging has always

Table 1.3 Packaging grouped by product packaged

1	Food (P, B, Gl, M, Pl)
2	Pharmaceuticals (P, B, Gl, M, Pl)
3	Cosmetics (P, B, Gl, M, Pl)
4	Chemicals and paints (Gl, M, Pl)
5	Gases (industrial and medical) (M)
6	Washing powders (P, B, Pl)
7	Cleaning products (P, B, Pl)
8	Appliances (electronics, photography) (B, Pl, W)
9	Textiles (P, B, Pl)
10	Vegetables and plants (P, B, Pl, W)

P:paper; Gl:glass; Pl:plastic; B:paperboard; M:metal; W:wood

been linked to the population's expectation of shelf-life. The industrial innovations relative to food processing and packaging during the 19th century, on the one hand, and the progress of the agricultural sector at the beginning of the 20th century in food supply (design of farm machinery, use of synthetic fertilizers, weedkillers, pesticides, etc.) on the other, have contributed to a complete breakdown of this dependence: most packaging nowadays is no longer so concerned with shelf-life but more with convenience, quality preservation and efficiency in transport and stacking. The latest applications are centred more and more on convenience, particularly when packaging is designed as a utilitarian item. Typical examples are the one-cup instant coffee packaged in a hermetic plastic bag wrapped in strong paperboard which turns into a cup when opened out, and the double-wall tin can, which allow the heating of prepared food without fire (the hydration of CaO produces heat).

For most foods, appropriate technologies in the handling, transport and packaging sectors have enabled their supply to be no longer dependent on harvest times and/or locations of origin.

By its colour and shape, printed information and/or selected material, packaging has become, more than ever before, a symbol. Besides the necessary functional aspect, it now possesses some emotional aspects capable of deeply influencing the saleability of the products. Emotional elements were already present in the packaging of the old rare and exotic goods some centuries ago. As sea routes were opened, the actual boom occurred after World War II with the emergence of supermarkets and self-service shops. In these shopping centres, pre-packaging is always the main requirement for an efficient and hygienic delivery. At the same time, packaging must induce the selling in addition to the promotion. Its colour, shape and printed information must be incentives for purchasing, and the packaging must give sufficient information to replace the former recommendation from the sales assistant.

Pre-packaging and the bundling of small units into bigger packagings (chocolate bars, nails, household articles) can be used to limit shoplifting. The big-looking packaging designed to mislead is not considered here; in the cosmetics sector, the packaging dimensions are often out of proportion to the actual packaged contents – the aim and the reasons for this in presenting rather expensive goods are mainly emotional.

The great expansion of self-service shops is shown in Table 1.4. Their part in the food retailing business is now about 85%. The fact that self-service shops are a modern and more efficient version of the former markets is not sufficient to explain the phenomenon. It seems more probable that the market system is well adapted to the Western way of living characterized by some sociological factors, namely a high standard of living, women's liberation, spreading urbanization, the short time available or devoted to cooking, high prices limiting domestic assistance, leisure opportunities offered to a larger proportion of the population, etc. These trends require not only a fast and efficient distribution chain (easy access, parking areas, well-ordered display of goods) but increase greatly the demand for pre-packaged and/or pre-cooked products.

Table 1.4 Expansion of self-service shops for foodstuffs in Belgium during the period 1949–1989 (situation on 1 January of each year)

Year	Number	Year	Number
1949	3	1969	2400
1950	10	1970	2800
1951	16	1971	3200
1952	23	1972	3550
1953	26	1973	3900
1954	31	1974	4050
1955	53	1975	4200
1956	87	1976	4300
1957	130	1977	4272
1958	166	1978	4278
1959	194	1979	4343
1960	288	1980	4200
1961	396	1981	4268
1962	583	1982	4300
1963	738	1983	4350
1964	1135	1984	4640
1965	1265	1985	4511
1966	1450	1986	4572
1967	1700	1987	4592
1968	2050	1988	–
		1989	4650

[Source: 'De ontwikkeling van de zelfbediening van voedingsmiddelen in België' in *Distributie vandaag*]

The interactions between evolution in the supply on the market, consumers' wishes, the general economic situation and employment, and the growth of single-person households are probably more complex than can be described here and are in keeping with a 'spiral' interdependence. However, it is a fact that this evolution could not possibly happen without the introduction of new packaging materials, the so-called synthetic materials or plastics. These new materials are generally suitable for fast, automatic pre-packaging of the largest variety of goods; they can also be combined with classical packaging materials (e.g. paper and aluminium foil) to improve their basic properties.

In reaction to competition (high labour cost of collection, transport, washing, etc.), single-trip packaging has been adopted for almost every kind of product. The food sector for example, has moved many glass packaging forms from multi-trip to single-trip packaging. A noticeable modification in composition, characteristics, weight and volume of household refuse is thus the logical impact of the extensive use of pre-packaging.

Plastics were not introduced initially to replace the classical packaging materials but to cope with an increasing demand. Later, these new packaging materials and techniques took precedence over the classical ones and/or new packaging were designed. We recognize in the evolution of packaging as described above the interdisciplinary dependency of discoveries, scientific and technical progress. Some important 'packaging dates' are summarized in Table 1.5.

b) Economic importance of packaging

Nowadays, with our way of living, marketing and distribution systems could not exist without packaging. More and more goods arrive pre-packaged on the market; since the introduction of supermarkets during the 1950s, packaging materials consumption in Europe has increased faster than the population. This is illustrated in Table 1.6 which shows consumption of various packaging materials during the period 1970–1989 in Belgium. The proportion of each material, expressed as a percentage of total consumption, can also be deduced from that table: it shows that in 1970 paper and board together represented about 46% (1977: 36%) of the consumed materials and, if glass was added, this percentage reached 68% (1970: 79%) of the total weight.

From the growth rates of these materials during the selected

Table 1.5 Important 'Packaging dates': an overview

3000 BC	Glass containers in Egypt
1500 BC	Paper, casks, baskets produced
300 AD	Glass containers common in Rome
1700	Paperboard boxes first produced
1809	Appert discovers preservation of food by heat
1810	Durand uses metal cans (soldered tin plates)
1846–1895	Mechanization of tin can-making up to first automatic machinery
1871	Albert Jones } issued patents for
1879	Oliver Long } corrugated board
1892	Mechanized glass bottle-making
1895	Wells Fargo accept 'cellular board boxes'
1914	Pridham decision
1925	Plastic era begins (PE)
1925–1930	Cellophane, PVC, PE, PS and Bakelite introduced in packaging
1935	First beer cans on the market
1936	Thermoforming used in France
1937	First extrusion blow-moulding
1946	PVdC (Saran) used as moisture barrier
1959	Aluminium used for cans
1964	Two piece Al impact extruded can developed in USA
1965	First tin-free steel can
1966–1968	Two-piece Drawn and Wall-ironed cans
1973	PVC and Acrylonitrile used for beverage bottles
1977	Retort pouches introduced
1977 –	Aseptics in flexible packing
	Modified-atmosphere packaging, controlled-atmosphere packaging
	Bag-in-box systems
	Vacuum packaging

period (see Table 1.7) we can determine which, among the latest packaging materials such as aluminium, corrugated fibreboard and plastics, really increase. Taking into account its low density (approx. 25 g for a litre bottle), the continuous extension of plastic consumption is even more significant than the

Table 1.6 The consumption of packaging materials and the production and deliveries of materials in Belgium (in 1000 tons) during the period 1970–1989

Year / Material	Subject	1970	1973	1975	1977	1979	1981	1983	1985	1987	1988	1989
Tinplate	Consumption	81	122	174	189	212*	261*	—	224	224	254	—
	Production	—	—	—	—	—	60	62	77	86	182	151
	Delivery	—	—	—	—	—	10	13	16	18	50	45
Aluminium	Consumption	12	13	13	14	*	*	—	—	—	—	—
	Production	—	—	—	—	—	169	218	147	138	72	71
	Delivery	—	—	—	—	—	73	95	65	59	32	28
Glass	Consumption	300	349	374	380	—	—	—	—	—	—	—
	Production	—	—	—	—	—	—	—	113	192	228	235
	Delivery	—	—	—	—	—	—	—	24	62	54	60
Paper	Consumption	134	162	125	116	152	182	123	137	132	—	—
	Production	—	—	—	—	—	96	97	107	89	95	100
	Delivery	—	—	—	—	—	70	68	73	63	67	68
Corrugated board	Consumption	157	249	210	215	221	255	292	341	364	—	—
	Production	—	—	—	—	—	217	229	291	281	305	266
	Delivery	—	—	—	—	—	150	164	190	190	235	179
Carton board	Consumption	65	63	64	53	65	60	73	73	98	—	—
	Production	—	—	—	—	—	60	96	72	99	106	208
	Delivery	—	—	—	—	—	41	65	43	59	58	123
Cellulose film	Consumption	6	8	6	6	**	**	**	**	**	**	**
Plastics	Consumption	95	101	94	110	119	112	123	125	155	—	—
	Production	—	—	—	—	—	112	123	125	155	174	203
	Delivery	—	—	—	—	—	55	62	67	72	82	90

*: Aluminium & tinplate together; **: included in plastics; —: not available
[Date concerning production and deliveries from National Institute of Statistics]

small increase of glass packaging in view of the high material density (min. 2.2 and max. 6.3 kg/dm^3) and of the weight of such packaging (approx. 500 g per 1 litre bottle). If we consider the data for Europe during the period 1973–1977, we notice first a stagnation: it could be the impact of influences such as the economic recession and zero population growth (cf. Table 1.6).

Similarly, this could lead us to conclude that a certain saturation has occurred in packaging consumption. It does not mean this sector has no future prospects; on the contrary, no industrial sector has been subject to so many improvements and to permanent adaptation, it follows the scientific findings so closely. The following innovations of the last 20 years prove it: the two-part aluminium can (now replaced by the two part tin-free steel can) for beer and soft drink packaging; the PVC bottle for mineral water; the pouches (bags or boxes made out of a combination including plastic, board and aluminium) for fruit juice and milk; the vacuum packaging for ground coffee (bag-in-box: a plastic bag in another bag, mostly in paper); the PET bottle for carbonated drinks and beers; developments in composite containers (plastics with metal can closures), etc.

It is not easy to isolate an industrial sector 'packaging'– besides the production of materials, a large number of other sectors are also concerned with packaging, such as the metal

Table 1.7 Growth rates of packaging materials' consumption in Belgium

Material	Growth rate (% per year)			
	1970–77	1977–81	1981–85	1985–87
Tinplate	19.0		>0*	0
Aluminium	2.4	5.7	—**	>0*
Glass	3.8	3.2	—**	—**
Paper	–1.9	14.2	–6.2	–1.8
Corrugated board	5.2	4.6	8.4	3.4
Carton board	–2.6	3.3	5.4	17
Plastics	2.2	0.5	2.9	12

>0: trend is positive; —**: data not available

industry manufacturing machinery for packaging shaping and packing, or the chemical industry designing adhesives and inks to print and make labels. Moreover, almost every firm has its own packaging department. Because of important ramifications in the industrial activities relative to production and consumption, it is also difficult to give figures for employment. Hence, the proportion of packaging in the gross national product is generally used instead: for Belgium it reached 4.4% (185) in 1980 and still remains today at that order of magnitude. The structure of packaging cost is also complex – this will become more evident with the description of the specific production processes, the materials and the energy consumption they induce. It is also interesting to take a closer look at the final price the consumer must pay for packaging. With the adoption of 'throwaway packaging', cost has become an integral part of the packaged product. Too often we underestimate the specific value and the value-added tax the consumer pays for the product's packaging, although its value as scrap is negligible as soon as it is discarded by the consumer.

When packaging is considered as part of the packaged product, its cost is quite different from the price that would be charged for packaging alone: moreover, for many products, empty packaging would not be delivered to the consumer! The distribution channel and the quantities are important here. The prices for a simple PE-paperboard packaging (pasteurized milk) and for a combination with Al-foil (UHT milk and fruit juices) amount to 8.30%, 10% and 13% respectively with regard to the total price. The price of the packaged content is not constant at all times, but packaging cost can vary either in the same proportion (especially when materials and energy prices are changing) or become dependent on competing production, recovery and returnable systems and the size of the order. The last-mentioned factor can have an important effect on the usual price of a packaging. This is shown in Table 1.8 where cost prices of various types of packaging are compared by material and the size of the order (220).

Table 1.8 Cost prices for different quantities of various types of packaging

Packaging	Volume of package (in litres)	unit cost per x thousand packs delivered (in BEF) x=2.5	=6	=10	=15	=50	=100
Glass bottle*	0.33	4.30	3.88	—	3.53	2.96	2.92
Euroglass bottle*	0.55	4.2	3.78	—	3.19	2.79	2.71
Light weight glass	0.33	2.1	1.96	—	1.76	1.65	1.62
Tin plate can	0.33	—	—	4.62	—	4.41	4.30
Al- can	0.35	—	—	4.62	—	4.41	4.30
Plastic tub	0.4	—	—	1.53	—	1.47	1.41
Tetrabrick	1.0	—	—	1.41	—	1.34	1.29

*multi-trip bottles
[Source: Paine (220)]

It appears that packaging cost can range from 2% to 25% of the selling price. It is quite surprising that packaging costs also reach 25% in the distribution sector (see Table 1.9). However, appropriate packaging should not always be considered as a 'cost'; as we shall see later, it always limits the damage and thus the losses of marketable goods. According to Grundke (71,72), these losses are attributed to the following factors: transport (54%), humidity (21%) and others (25%); packaging in reality shrinks costs when it enables a functional handling of the goods from the production through to retailing (stacking on pallets during warehousing, transport and distribution).

A relevant thorough survey (85) highlights the fact that collective packaging which meets standard packaging gauges can result in savings. As the consumer's carrying strength and consumption rate are limited, optimal enlarging must be sought; a collective packaging including 12 boxes of UHT milk (approx. weight 13 kg) is indeed economical because this unit allows the best stacking conditions for transport (dimensions within 1000 mm × 1200 mm ISO standard for pallets; see also page 56), it saves space and areas during distribution and is most convenient for mechanical handling (trucks, conveyor belts, etc.) resulting also in savings on labour costs. Figure 1.5 illustrates the part of packaging cost depending on the scale

Table 1.9 Price structure in the distribution sector

Administration	35%
Warehousing and handling	35%
Transport	5%
Packaging	25%

[Source: Jönson (85)]

increase (regrouping of units) in the distribution sector. These data emphasize the economic importance of packaging; collaboration between producers, wholesalers and retailers may lead to a rationalization justified also in international trading, that might lead to minimal material consumption (e.g. by using recyclable paperboard packaging) and lower pollution of the environment (e.g. more efficient transport results in lower fuel consumption and air pollution).

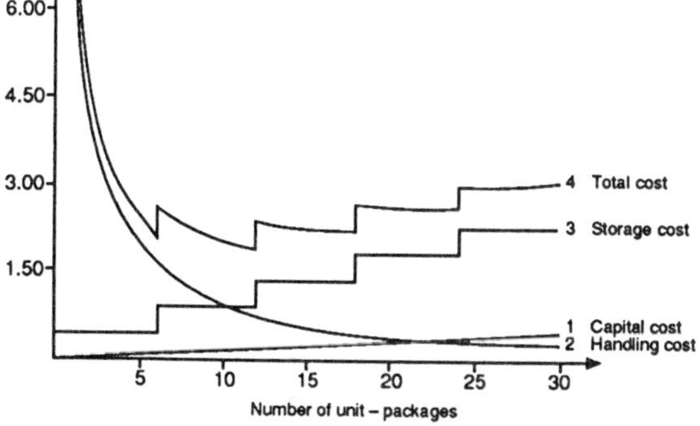

Figure 1.5 Cost price structure of packaging as a function of unit sizes [Source: Jönson (85)]

3 THE FUNCTIONS OF PACKAGING

Packaging is indispensable to the diverse interested parties through whose hands every marketable product passes, at least from its creation through to its use. After primary packaging, the goods in their individual packaging are collated for transport and sale. During this period, starting with the packer and ending with the consumer, the packaging must offer protection against any influence from the surroundings or arising from transport, warehousing, distribution and use. At every stage of the product's life, its packaging must ensure the preservation of quantity and quality and contribute to the saleability and/or price. Some of these properties that packaging must feature are indicated in Table 1.10.

Interest in the ecological aspects of packaging materials have been initiated by consumers and their organizations. Most sectors have in the past paid little or no attention to the ecological aspects of packaging: this situation was likely to continue as long as the relevant financial and economic implications stayed low. Nowadays, manufacturers and retailers demonstrate clearly their responsibility to environmental hygiene: there is no doubt that the 'green image' has become part of the market's philosophy.

Table 1.10 emphasizes the immediate interest: between the parties, there always exists a certain reciprocal dependence. The market research that is always involved in the introduction of a new product and its packaging is not dealt with here. We must, however, mention Kotler and Mickwitz's theory (95,119) describing clearly the importance of packaging for the saleability and/or the return of a specific product. When the sales of a product reach a saturation phase, packaging can significantly influence the maintenance of a favourable sales increase (see Figure 1.6).

Packaging becomes part of domestic refuse as soon as the consumer separates it finally from the product.

Until the late 1980s, it was mainly the consumer who was directly confronted with used packaging and its disposal. Separate collection of materials initially needs surface and/or

Table 1.10 Packaging requirements needed by various interested parties from production through distribution to consumption

	Manu-facturer	Packer	Transport sector	Ware-housing (stacking)	Distribution	Consumer
Packaging cost	x	x			x	x
Nature of packaging material:						
a. relative to physical, chemical and mechanical properties	x	x	x	x	x	x
b. relative to preserving the quality of the product		x		x	x	x
Feasibility of automatic packaging		x				
Good printability (labelling)		x				
Information about the product					x	x
Feasibility of stacking (palletization)			x	x	x	
Appearance and emotional elements					x	(x)
Ease of handling			(x)	(x)	(x)	x
Goods coding by means of packaging	(x)	x	x	x	x	
Ecological aspects of packaging	x				x	x

x: major interest; (x): minor interest

space at the consumer's dwelling and is based upon the willing cooperation of a large part of the population. As far as the quality of life is concerned, it is in everybody's interest that the utmost is done to collect and treat the refuse efficiently. In this

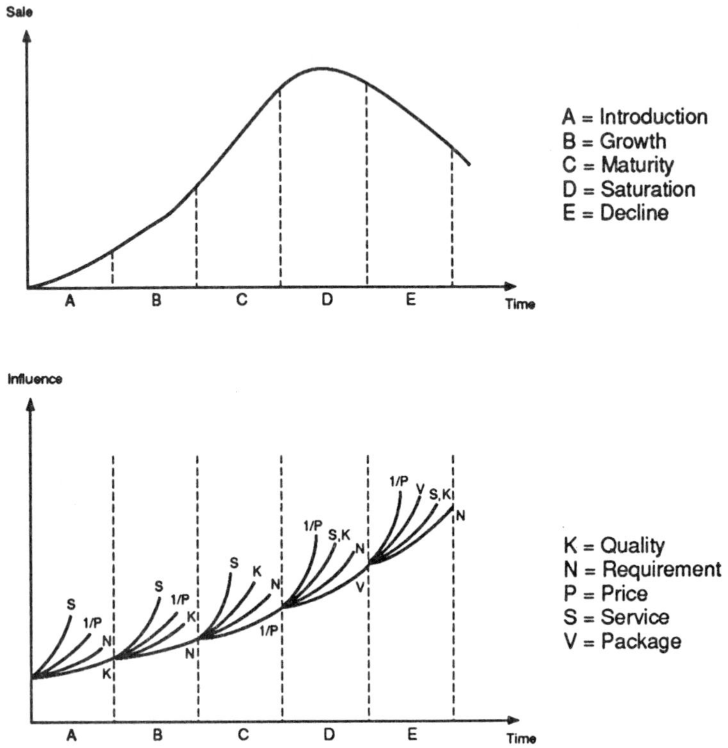

Figure 1.6 The normal turnover of a product as a function of time
[Source: Kotler and Mickwitz (95)]

respect, an 'environment tax' imposed by local authorities may be justifiable.

Table 1.10 also points out that packaging must cope with three main functions:

1. a protective function to preserve the quality of the product,
2. a commercial function to influence the saleability of the product,
3. a control function to simplify cataloguing and automatic data processing (e.g. for purchasing and stock control) throughout the distribution chain.

A fourth function is coming into force, namely a material

identification function: packaging (especially plastic) will carry information (e.g. a colour code or embossing) for identification of the different materials. This would enable a separation of household refuse into its components and subcomponents (sorting by plastic groups) and could result in better management of waste disposal treatment and materials recovery.

a) Protective function of packaging

The protective function is the most important one. Dealing with a 'protective function' implies that we have deep knowledge of all aspects related to the packaged product and can identify what should be 'protected'. In this respect, we must be able to specify and determine accurately the quality of the product, we must know every hazard that could provoke alterations and thus modify the quality, and we must also know every factor which could be present in the external surroundings (see Figure 1.1) during manufacture, packaging, transport, warehousing and distribution and which might affect that quality. Taking all this into account, it should be possible to design an ideal package ensuring preservation of the quality until consumption and/or use.

The ambient parameters, such as temperature, humidity, atmospheric pressure and light intensity, can vary during transport and warehousing: the surroundings can also be quite different in the consumer's country, in the country of origin or during manufacture and packing. However, such parameters can generally be measured with simple instruments and they can be expressed in physical units and reproduced in laboratories; the simulation conditions to test these 'quality parameters' of packaging are relatively easy to define.

It is more difficult to describe accurately and to quantify the potential mechanical hazards that packaging is subject to during handling (packing, loading, unloading) and transport in order to truly reproduce them in the laboratory. For many types of transport, packaging protocols and standards relative to test procedures are still being studied. The shocks, vibrations and movements that packages suffer during transportation vary with the transport mode (truck, train, airplane, ship); moreover the forces acting on the package inside a single vehicle may depend on its location. A package is not submitted to the same stresses when placed along the side of a lorry as when situated in the middle, the front, the rear or at the bottom or top of a stack.

The packaged goods can also be vibrated inside their packaging, either relative to each other and/or to the walls of the package. If a resonant frequency is reached due to external circumstances, both product and packaging will be highly stressed. Such potentially damaging phenomena must also be taken into account when devising test procedures and instructions for the transport and handling of packages. In addition to the mechanical and physico-chemical influences against which packaging must provide protection (and which it must itself resist), we must also consider biological factors such as micro-organisms (e.g. bacteria and moulds) and insects capable of either damaging the package or entering it through small openings (e.g. pinholes and the loose weft of textiles), factors which can alter the quantity and the quality of the contents. For food packaging, special attention must be paid to this aspect (see also page 201 and following). Rodents can also damage a package with a resulting loss of quality. The main interactions between packaging and the external surroundings are summarized in Table 1.11. The interactions between a package in contact with its content are considered in detail in Chapter VI. All this goes to show that it is quite complex to design and make an ideal package (222).

Calculations for the design of packaging able to provide a specified degree of protection against the ambient atmospheric factors are based on physical laws. A wise choice of the factors to be protected against avoids over-packaging and over-protection and thus the squandering of packaging materials. This choice of factors is based on a knowledge of the qualitative characteristics of the packaged product. In this connection, the 'quality' concept is not always easy to define or specify: quality can be expressed as a percentage if it is possible to indicate the presence of a particular characteristic and compare it to an ideal value fixed in advance, e.g. the drop in vitamin content of foodstuffs coming into contact with atmospheric oxygen can be used to estimate the quality reduction.

Table 1.11 Packaging protective function against external factors

External factors	Functional elements and required properties for packaging and packaging material
Mechanical shocks,vibrations and compressive loads	Resistance (brittleness) Bending strength Shock absorption Tensile strength (elasticity) Compressive strength (resistance to stacking) Seal resistance
Gases (oxygen, nitrogen, carbon dioxide), water vapour, odours such as aromas and flavours and their mixture	Permeability Porosity and density
Liquids (water, oil, etc.)	Solubility, absorption
Light intensity and radiation energy (ultraviolet light, visible and infrared (IR) light)	Light transmission characteristics Light absorption Reflection (brilliancy)
Thermal content and temperature	Thermal conductivity
Biological factors	Resistance or sensitivity

For most products, each consumer will have recourse to a total evaluation based on some quality parameters such as appearance, colour, flavour, weight, fragility, utility, presence of a label referring to the test procedures performed, price, promotion, etc. Thus, quality is to be considered in every sense of the word, that is to say as the measure of compliance (or respect) with a certain number of desirable features (70). As these can vary according to the social class and the population group, or may be linked with customs, the concept of 'quality' has a social implication related to the standard of living. Hence 'packaging' can also be considered as a quality criterion (see also 'Emotional and pertinent elements' on page 50).

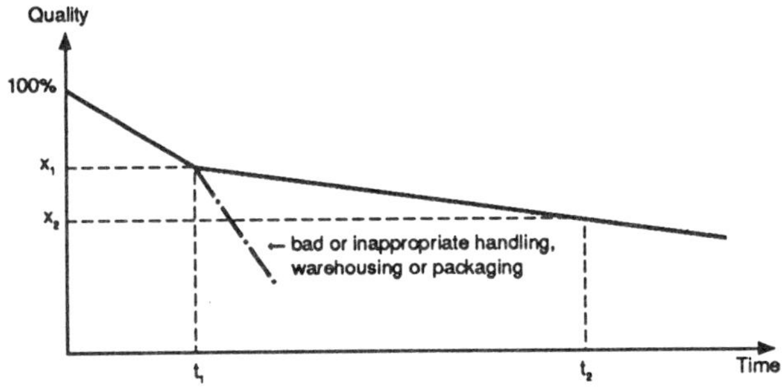

x_1 maximum attainable quality level
x_2 predefined limit (accepted quality decrease)
t_1 period of manufacture, handling and packaging
t_2 period guaranteed or deadline for use (expiry date)

Figure 1.7 Quality evolution of packaged foodstuffs as a function of the conservation period.

It is generally accepted that the quality of a product decreases progressively (see Figure 1.7). Some exceptions do exist e.g. wine, processed meat, fresh fruits, some wooden materials, and lead (by diminution of radioactive elements). In these cases, the quality as a function of time can momentarily reach a maximum before decreasing.

Protective function of packaging against mechanical influences

Preservation of package integrity depends mainly on the mechanical properties of the packaging form used. Packaging must resist not only external stresses but also internal ones caused by the packaged goods themselves. The choice of adequate packaging is thus partly determined by the nature of the product and partly by the circumstances required by the packaging operation, the condition of transport and handling, and the specific purpose (e.g. simple wrapping of a utilitarian article) (65). The resistance and/or fragility of the unit packages determine the type of transport packaging selected. Product parameters are mainly stiffness, fragility and temperature. Temperature variations have an influence on volume; they can induce a variation of the vapour pressure above a liquid apt to cause either an over-pressure or a vacuum in the package. For this reason, a bottle should never be filled to the brim, and a metal can full of beer or soft drink is less compressible than an empty one. The mechanical influences that packaging must resist produce different stresses: the ways in which these stresses are applied are illustrated in Figure 1.8. Test procedures aim to simulate as closely as possible the actual conditions; packaging materials (paper and board, plastic films, metal foils, etc.) and/or packaging forms are tested according to the specification's requirements in defined conditions of temperature and humidity (moisture content).

For most materials, tensile or elasticity tests performed at various temperatures give the main information on their efficiency. Under tension the length will first increase in direct

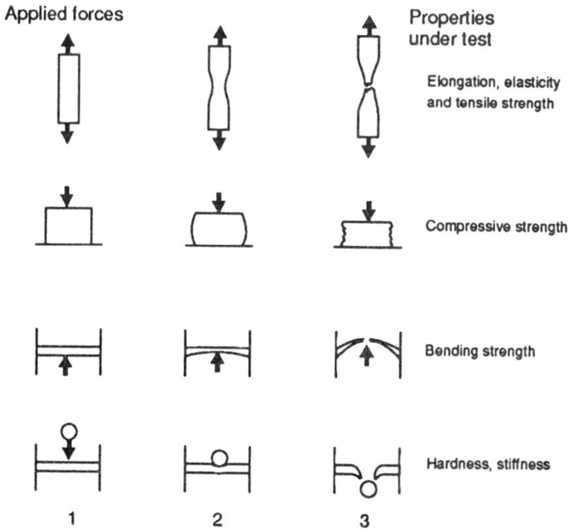

Figure 1.8 Schematic stress diagrams to test the mechanical properties of packaging wares.
The arrows indicate the direction of applied forces. Three steps are illustrated: (1) at the beginning; (2) during deformation; (3) at the rupture of the test sample.

proportion with the applied load, while the cross section will shrink. Within this linear area (see Figure 1.9), deformation remains elastic: when the load is removed, the strip returns to its initial length. The ratio between elongation and applied tensile stress is called the modulus of elasticity (Young's modulus). Once past the limit of elasticity, the length increases faster than the applied load. Table 1.12 indicates for various packaging materials the maximum tensile stress per unit of cross section before rupture. This value depends mainly on the material, its molecular structure and the temperature (cf. Figure 1.9). For instance, plastics feature an important variation of this maximum tension stress near the glass transition temperature (see Chapter IV).

A high tensile strength is required of material for bags: they must not stretch under the weight of their load. Modification of the cross section as a function of the length is also influenced by the degree of crystallization of plastic materials; as we shall see later on, that feature of plastics can be modified by changing the reaction parameters (temperature, pressure and catalysts). The Poisson coefficient given in Table 1.12 is the expression of this relationship.

Crushing is generally caused by stacking loads during warehousing or transportation. When moved, packages are submitted to vibrations which cause supplementary accelerations in three dimensions. Hence, the primary packages must resist forces bigger than those calculated from expected stack heights. Forces originating from vibrations or from all sorts of impacts (drops, shocks) should be absorbed by the bending properties of the materials used or by using cushioning materials. The effect that any impact will have on an article (package or packaged product) is also related to its fragility – the ease with which it can be damaged. The fragility factor, or G-factor, or impact–load–product is usually expressed in units of earth gravity (g). A product with a G-Factor of 20 means that it will resist, without damage, impacts equal to 20 times its own weight. Cushioning materials must have the property to decelerate the products to levels below their fragility. The resistance

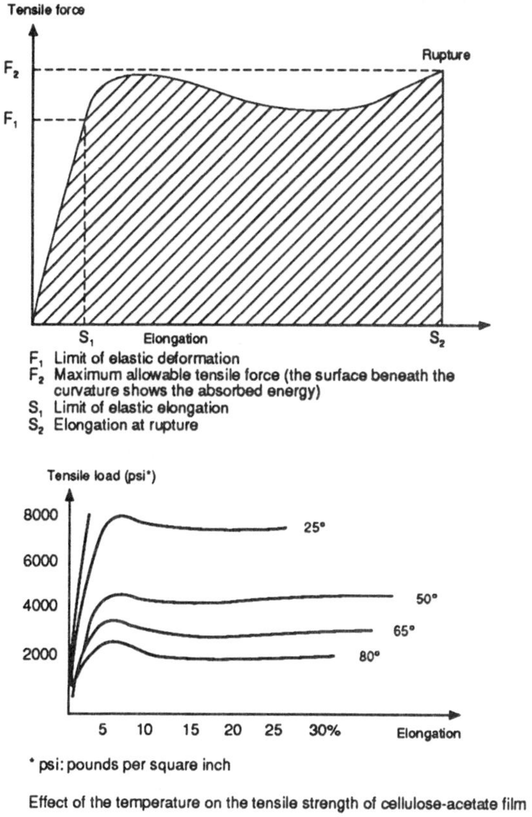

Figure 1.9 Typical evolution of the percentage variation in length a test specimen undergoes as a function of the tensile load (Newton per unit of cross section)
* psi: pounds per square inch
F_1: Limit of elastic deformation; F_2: Maximum allowable tensile force (the surface beneath the curvature shows the absorbed energy); S_1: Limit of elastic elongation; S_2: Elongation at rupture
[Source: Carswell T S, Nason H K (29)]

of packaging to stacking, vibration and shocks forms the most important part of its quality label.

Resistance to stacking height is tested by the so-called compression test: the packages are loaded in order to simulate the real conditions. Resistance to vibrations is tested by fixing the packages to the table of a vibration machine which will produce

Table 1.12 Comparison of mechanical properties featured by various packaging materials

Material	Young's modulus (N/m^2)	Poisson coefficient	Maximum tensile strength psi*	Density (kg/dm^3)
Iron	9 10^{10}	0.27	10.340	7.8
Aluminium	7 10^{10}	0.33	6.206	2.7
Glass	6 10^{10}	0.23	6.890	2.5
Tin	4 10^{10}	0.35	2.760	7.3
PS	3.4 10^{9}	0.33	4.140	1.05–1.07
PET	2.8 10^{9}	—	7.685	1.38–1.41
PVC	2.8 10^{9}	0.4	7.975	1.38–1.41
Nylon (66)	2.0 10^{9}	0.3	6.890	1.13
PP	0.8 10^{9}	—	7.350	0.85–0.92
HDPE	0.9 10^{9}	—	5.200	0.94–0.98
LDPE	2.4 10^{8}	0.38	1.670	0.89–0.93
Rubber	2.0 10^{6}	0.49	1.380	1.6

* psi: pounds per square inch; 1 psi = 6895 N/m^2 (Pascal)

[Source: *Encyclopedie der materialenkennis* (184); *Concise Encyclopedia of Polymer Science and Engineering* (211)]

horizontally or vertically-oriented accelerations. Rotating machines will move the table back and forwards or up and down over a distance of 2.5 cm: adjusting the speed from 30 to 300 revolutions per minute, accelerations from 0.1 to 1 g can be simulated.

More sophisticated apparatus allows a scan from 3 to 200 Hz, the displacement being much lower (ca. 250 μm). With the latter, resonances can be detected as well. Both methods of testing will simulate vibrations and accelerations found in different transport modes as indicated in Table 1.13.

Table 1.13 Vibration characteristics of the main transport systems

Transport system	Origin of vibration	Characteristic frequency (Hz)	Acceleration (g)
Rail	Vertical suspension	2–7	0.50
	Lateral suspension	0.75–2	0.75
	Structure	50–70	0.25
Road	Suspension	2–4–7	0.50
	Vibration of non-suspended masses	10–20	0.25
	Structure	50–100	0.25
Sea	Pitching, rolling	—	1.20–1.80
	Engine vibration	100	low
Air	Take-off, landing	100–200	high

[Source: Paine (222)]

Possible stresses from shocks and other dynamic impacts during transport and handling can be simulated by the so-called drop tests. In these tests, packages are dropped from a norm-specified height on to a thick steel plate or concrete floor. The drop height is adjusted from 0.3 to 1.2 m depending on the weight, ranging respectively from 400 to 10 kg. Here shock absorption, damping capacity or the strength of the packaging are tested.

Taking these values into account with the specific weights of the materials, it is possible to design ideal packaging with the required strength and damping characteristics. Computer-aided design techniques can elaborate prototypes offering the most desirable material with regard to mechanical properties and price. There are generally marks on transport packaging pointing out fragility, sensitivity to temperature, inflammability, and the danger or damage incorrect handling can provoke; often preventive actions are also suggested. Figure 1.10 reproduces some of these symbols.

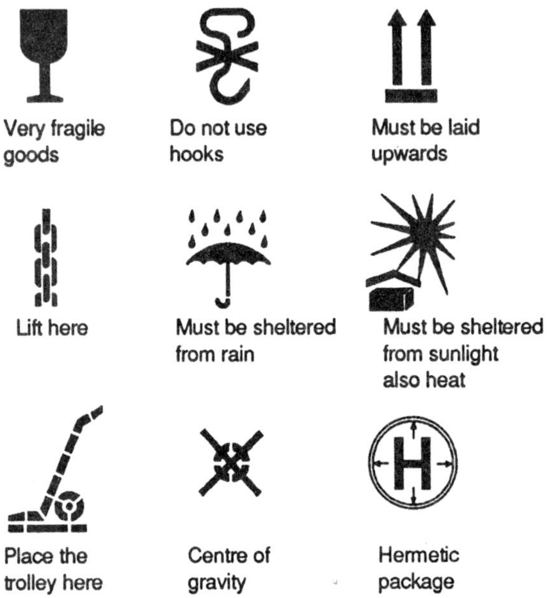

Figure 1.10 Standard symbols used to label transport packaging.

Protective function of packaging against the influence of light

Light is an electromagnetic radiation; its energy quantum depends on frequency according to:

$E = h\upsilon$

where E = energy
h = Planck's constant
υ = frequency of the light

Consequently, the higher the frequency, the more energetic the light. Wave length (λ) can be calculated from the specified frequency:

$$\lambda = \frac{c}{\upsilon}$$

where c = speed of light 3.10^8 m/sec

Materials can either reflect, filter, or absorb light. Generally all three phenomena are present together: absorbed light can cause chemical reactions, and/or accelerate catalysis. Fading of materials under the influence of light is one of the most striking effects. Foodstuffs contain many photo-sensitive components such as vitamins (vitamin B2 and vitamin C), pigments (β-carotene), proteins, fats and oils. Generally light accelerates oxidation and/or decomposition by atmospheric oxygen. Oxidation of fats (becoming rancid and/or souring) (89) not only lowers the market value and/or the nutritional value (sometimes associated with the formation of toxic products), but is also involved in the decomposition and loss of oil-soluble vitamins such as vitamins E and A. Many solid foodstuffs, such as bread or smoked meats, are protected against the influence of light by the crust and/or by their own opacity; light will be absorbed almost completely in an external layer. Liquids do not have such protection because they are always agitated by convection and are therefore exposed continuously to the action of light.

For example, in a test at 4°C, milk in various types of packaging was exposed to constant illumination: a sharp drop in vitamin C and vitamin B (riboflavin) content occurred. The results are summarized in Table 1.14.

Karel (89) pointed out that illumination in warehouses and supermarkets can be four to five times more intense than in this experiment; consequently the loss can be even higher. At higher temperatures, the convection flows are more intense resulting in a faster quality decrease.

The nature of the protective function of packaging is twofold: it should minimize the light passing through the packaging, and also prevent elements (atmospheric oxygen, for instance), needed for oxidation reactions accelerated by light, passing through.

Table 1.14 Effect of package type upon the fall (%) in vitamin content of milk after 24 hours at 4°C under permanent illumination (light intensity 1076 lux)

Vitamin	Normal content (mg/l)	Packaging Paperboard	Transparent plastic bag	PE bottle	Black PE bottle
Vitamin C	10.0	23.6	86.2	90.9	13.0
Vitamin B2	1.7	7.1	11.1	22.3	0.0

[Source: Karel (89)]

The fraction of incident light passing through the packaging surface follows Lambert's law:

$I = I_0 . e^{-kx}$ where I_0 = intensity of incident light
I = intensity of transmitted light
x = thickness of packaging material
k = absorption ratio (related to the wave length and material) (cf. Figure 1.11)

$I/I_0 \times 100$ is the light transmission (expressed in per cent); its variation is a logarithmic function of the thickness. Conversely, light absorption (or fading) worked out by $\log I/I_0 = k'x$ where $k' = 0.4343$ (conversion from Naperian to decimal logarithms) and varies linearly with the distance. Knowing the absorption ratio, it is possible to determine the thickness of a package, as a function of a specified light absorption, in order to meet a required protection level for the product. Small absorption ratios involve thick walls to reach the required result. Taking into account both the mechanical properties of some materials and the light absorption ratio of others, we can still obtain good results with thin walls for the package (low material consumption) by combining different films. Three- to five-layer laminates can be used to make packages which feature sufficient tensile strength and provide a good barrier to both air and light penetration. For a three-layer packaging, (see also Figure 1.11):

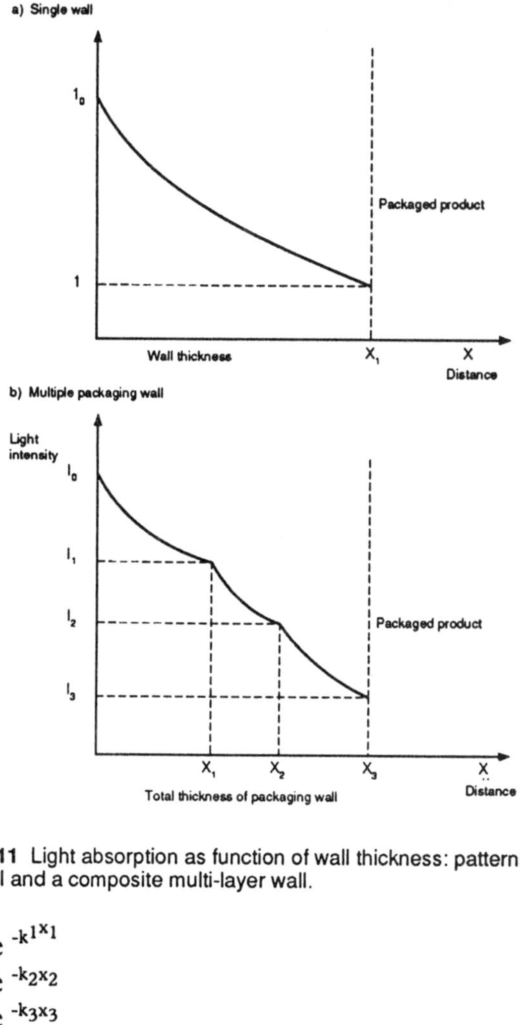

Figure 1.11 Light absorption as function of wall thickness: patterns in case of a single wall and a composite multi-layer wall.

$I_1 = I_0\, e^{-k_1x_1}$
$I_2 = I_1\, e^{-k_2x_2}$
$I_3 = I_2\, e^{-k_3x_3}$

Hence $I_3 = I_0\, e^{-(k_1x_1 + k_2x_2 + k_3x_3)}$

and the total absorption is

$\log I_0/I_3 = k'_1 x_1 + k'_2 x_2 + k'_3 x_3$
where k'_1, k'_2, k'_3, x_1, x_2 and x_3 are respectively the absorption coefficients and the thicknesses of the three components

The quanta of light of various wavelengths absorbed by the packaging can be worked out by means of these formulae. Then the penetrating light will not necessarily be completely absorbed by the packaged product: some can always be reflected.

Karel mentions (89) that the light absorbed by the packaged product can be calculated from:

$$I_{abs} = I_0 . Trp. \frac{1-Rf}{1-Rf.Rp}$$

where I_{abs} = light absorbed by the packaged product
I_0 = intensity of incident light, measured at the external side of the package
Trp = transmission, i.e fraction of light passing through the package
Rp = fraction of incident light reflected by the package
Rf = fraction of penetrating light reflected by the packaged product

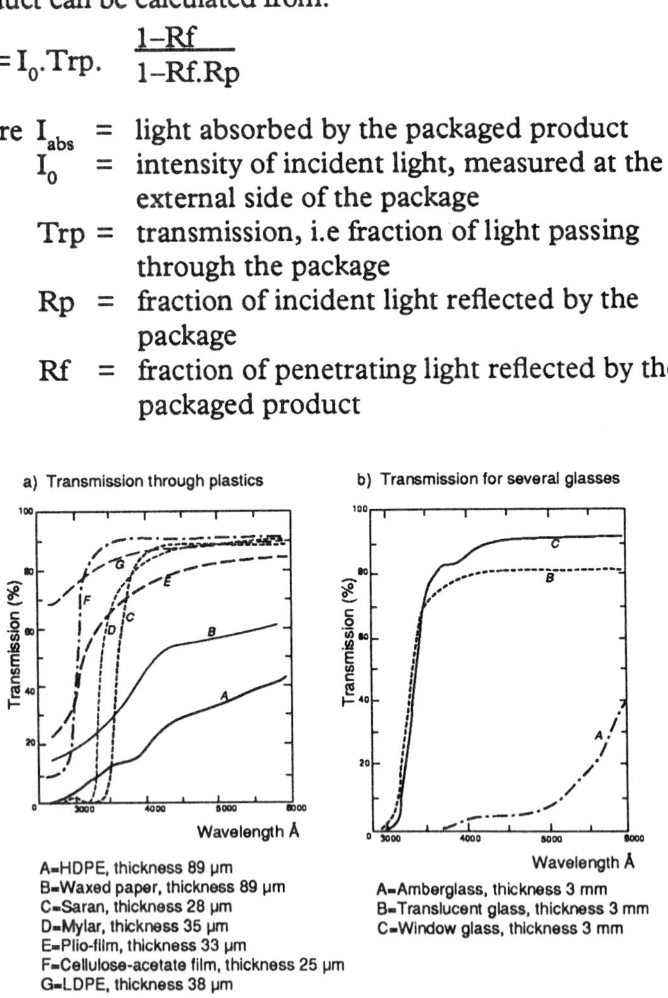

Figure 1.12 Light transmission curves for some food packaging materials as a function of wavelength

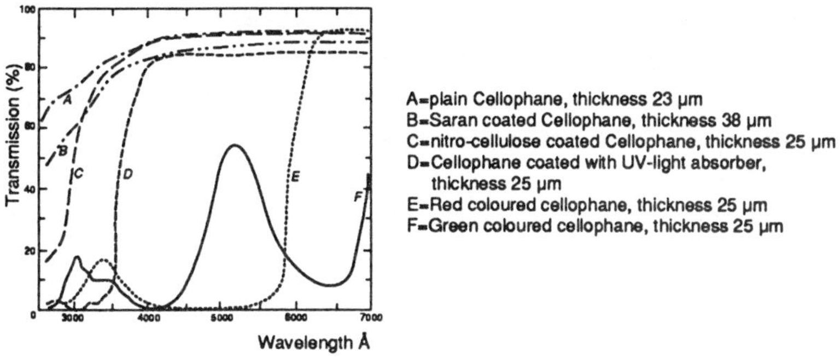

Figure 1.13 Light transmission curves for different types of Cellophane as a function of wavelength

Taking into account the various characteristics of the packaging and the packaged product, it is possible to estimate the fall in quality with time. In practice, we work conversely and calculate the allowable limits for light penetration to cope with a predetermined acceptable fall in quality within a defined conservation period (e.g. a few days for meat, three to six months for milk). With these data, the most convenient material combination can be determined (182). Some transmission curves of packaging materials are shown in Figures 1.12 and 1.13.

Figure 1.12(a) shows that, taking the thickness into account, HD polyethylene gives better protection against UV light than saran, mylar, cellulose acetate, LD polyethylene and waxed paper. Figure 1.12(b) shows that, as expected, less light passes through coloured glass (brown, green) than through plain glass but the transmission of UV light is nearly the same. Figure 1.13 demonstrates that coloured films have the required absorption over a wide spectral range and offer better protection against the influence of light.

Protective function of packaging against gases and vapours

Many products made of metals, plastics, foodstuffs etc. are affected by atmospheric oxygen and undergo modifications resulting mainly in a degradation of quality. The rusting of iron is a slow oxidation process, accelerated by water and high temperatures. Other metals used in packaging such as tin and aluminium, under the influence of oxygen, form an oxide layer which can offer protection against further corrosion. Some plastics become friable under the influence of atmospheric oxygen and exposure to solar radiation; hence, anti-oxidants are added during their processing. Many food components such as fats and oils, albumen and vitamins, in the presence of oxygen, undergo chemical reactions which modify their appearance (e.g. meat and peeled potatoes turn brown), their taste (fats become rancid) and their odour. An abundance of literature (89, 96, 97, 112, 138, 152, 153, 154) exists on the fall in vitamin content (vitamins C, E and A) in foodstuffs occurring not only during their processing but also during storage. These oxidation reactions may also be the indirect result of oxidation: e.g. fatty acid peroxide created from fat, oxygen and light, will serve as a catalyst and promote the oxidation of many albumens and amino acid components (89). Generally these oxidation reactions require atmospheric oxygen, light and water and are induced by the presence of enzymes in the foods.

Often water loss or gain results in a lowering of food quality: the relevant protection is related to the water activity a_w of the products:

$$a_w = \frac{p}{p_0}$$

where p = vapour pressure induced by the product
p_0 = vapour pressure measured above pure water

For example, meat with an a_w above 0.95 placed in a relative humidity (rh) of 75% will lose moisture and dry out, while biscuits with an a_w of 0.3 will absorb moisture (see Figure 1.14). Moulds may appear on foodstuffs if water activity is over 0.7; biscuits lose their crispness at a water activity above 0.35, sugar may crystallize or dissolve and the texture of confectionery may become organoleptically unacceptable when a_w is respectively too low or too high. Besides its dependence on temperature and humidity content, the shelf-life of fruits and vegetables is also significantly influenced by the oxygen/carbon dioxide ratio of the surrounding air; fresh unprocessed products 'breathe' during their shelf-life producing carbon dioxide and water at the expense of atmospheric oxygen. If oxygen is missing, or if the supply of water and carbon dioxide is not sufficient, rotting begins. The ratio between the amount of carbon dioxide produced and the amount of oxygen consumed is called the respiratory quotient:

$$\text{RQ (respiratory quotient)} = \frac{CO_2 \text{ (produced)}}{O_2 \text{ (consumed)}}$$

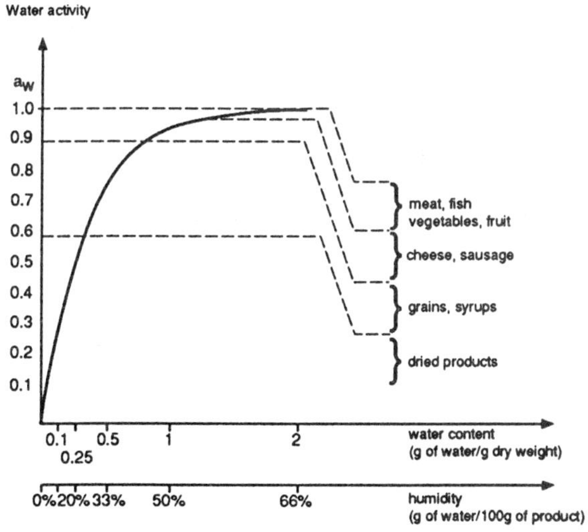

Figure 1.14 Water activity as a function of the water content of various foodstuffs

For carbohydrates (sugars, honey products) RQ = 1; the 'biological combustion' of products rich in albumen consumes more oxygen than it produces carbon dioxide (RQ < 1); conversely, during rotting, oxygen is removed from materials rich in oxygen (sulphates, nitrates, etc.) – hence, RQ > 1.

From this short summary it is clear that, as far as quality preservation is concerned, packaging has a great influence on the exchange of gases and vapours between the external surroundings and the internal package atmosphere.

In order to be able to estimate the shelf-life of products (shelf-life prediction), we must have accurate data about the permeability of packaging materials to gases such as oxygen, nitrogen, carbon dioxide and to vapours such as water vapour, aromas, etc. (77, 96, 97). Transmission of gases and vapours through packaging material as a function of time can be expressed by the equation:

$$F = \frac{dM}{dt} = k \; \frac{D}{x} \, . \, S(p_1 - p_2)$$

where dM/dt = flux F, mass of gas (vapour) transmitted per unit of time (in gram/cm^3 or mole/day)
k = solubility constant of the gas in the packaging surface (mole/cm^3/bar)
D = diffusion constant of the gas in the packaging material (cm^2/day)
x = thickness of the packaging material (in μm)
S = packaging area (in m^2)
$\Delta p = p_1 - p_2$ = Difference in the partial gas (vapour) pressure between internal and external condition (in mm Hg, or in atmosphere, or in bar) (cf. Figure 1.15)

The product k.D is called the permeability constant B

$$B = \frac{F.x}{S.\Delta p}$$

Thus its dimension is $\dfrac{\text{mole}}{\text{day}} \cdot \dfrac{\mu\text{m}}{\text{m}^2.\text{bar}}$

The permeability constant is defined as the amount of gas (vapour) diffusing per unit of time through a packaging film, the thickness of which is one unit related to one unit of area

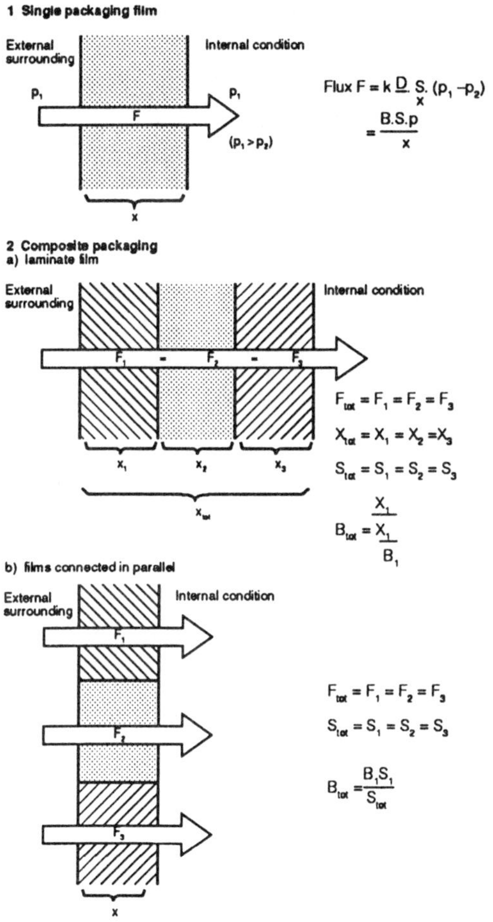

Figure 1.15 Pattern of the gas (vapour) transmission through packaging materials

and partial pressure drop. Quite heterogeneous permeability values are published in the literature: on the one hand, the amount of gas (vapour) that can diffuse is expressed either in weight (gram) or in volume (ml, cm^3, $inch^3$) or in mole (gram/molecular mass), and on the other hand, the partial pressure is indicated either in atmosphere or in mm of mercury or in bar.

Conversion between units follows the Boyle-Gay Lussac relationship.

$$P.V = \frac{x}{MM} .R.T$$

where V = volume in litre
P = partial pressure in bar
x = weight in gram
MM = molecular mass of the diffusing gas
x/MM = number of mole of the particular gas
R = general gas constant (0.082)
T = absolute temperature (degrees Kelvin = degrees Celsius + 273)

So the volume of 16 mg oxygen at 27°C and at 1 bar is equal to

$$V = \frac{16.10^{-3}}{32} . 0.082 . 300 \text{ litre}$$
$$= 12.3 . 10^{-3} \text{ litre}$$
$$= 12.3 \text{ ml}$$

As thickness unit we use either the 'mil' (25.4 μm) or 100 μm. A permeability constant of 3800 cm^3.100 μm/day.m^2.bar at 30°C for oxygen (cf. Table 1.15) means that through a 100 μm thick film and a surface of 1 m^2, at a temperature of 30°C, 3.8 litre oxygen is diffused each day. If the film is only 20 μm thick, it will allow the penetration of five times as much oxygen, i.e. 19 litres.

According to the formula, we can convert the oxygen quantity, expressed in litres, into weight:

$$x = \frac{19 \times 32}{0.082 \times 303} = 24.5 \text{ g}$$

Permeability of glass and metals to gases is negligible at ambient temperatures although metal and glass packages can leak at the welded joints and through closures. Predictions about the preservation of the product quality based on gas exchanges are mainly valid for plastic films. Permeability to oxygen, nitrogen and carbon dioxide is given for various plastic films in Tables 1.15 and 1.16.

Table 1.15 Permeability to oxygen at 30°C of various packaging materials

Material	Permeability constant expressed in	
	$\frac{cm^3.mil}{day.m^2.bar}$	$\frac{cm^3.\ 100\ \mu m}{day.m^2.bar}$
LDPE	6000–15000	1500–3800
HDPE	1500–3000	380–750
Saran (PVC-PVdC)	10–350	3–90
Cellophane	20–5000 (2)	5–1300
Teflon	25–100	6–25
Mylar (polyester)	50–100	13–25
Poly-trifluoro-chloro-ethylene	50–1000 (1)	13–250
Cellulose acetate	1000–3000	250–750
PE coated with paper	100–15000 (3)	25–3800
PP (oriented)	2450	622
PP (n. oriented)	3670	933
Aluminium-laminates	0	0
Plastic laminates	10–400	3–100
Nylon 66	55–244 (2)	14–62
PS	3980	1010

(1) depending on the amount of plasticizer; (2) depending on the moisture content; (3) depending on the coating material
[Source: Karel and Heiss (89,77); Bakker (188)]

In the case of plastics, the permeability to gases depends considerably on the temperature. As we will see when we discuss the shrinkage properties of plastic films, plastics have a rubber structure in the temperature range between the so-called glass transition temperature 'tg' and the melting point; at temperatures lower than tg the structure is glass-like (partly crystalline, partly amorphous) – permeability decreases when the molecules are better ordered (i.e. a higher degree of crystallization). As with natural rubber, plastics within the rubbery field have a higher permeability to gases (149). Permeability to water vapour of various plastic films is given in Table 1.16.

The moisture content of water-absorbing packaging materials such as regenerated cellulose, cellulose acetate and polyvinylalcohol can have a great influence on the permeability to gases. The water acts as a 'plasticizer' and water uptake causes the plastic film to swell, increasing significantly the permeability to gases and vapours.

Table 1.16 Comparison of the permeability constant B, relative to various gases at 20°C, of some plastic films $\left(\frac{cm^3.100\ \mu m}{day.m^2.bar}\right)$

Material	Nitrogen (B_{N_2})	Oxygen (B_{O_2})	Carbon dioxide (B_{CO_2})	$\frac{B_{CO_2}}{B_{O_2}}$
Saran (PVC-PVdC)	1	3.3	19	5.7
Nylon 6	6.3	25.4	102	4
Mylar (polyester)	5	20	66	3.3
LDPE	890	3050	17800	5.8
HDPE	178	508	2540	5
Rubber	5080	15200	89000	5.8
PP (non-oriented)	142	937	2300	2.5
PP (oriented)	80	622	2160	3.5
PS	—	1014	3350	3.3
PET (oriented)	3–4	24–32	60–100	2.5-3.1
PVC (plain)	10–25	20–80	40–200	2–2.5
Cellophane	2–6	0.8–2	1.5–24	1.8–12
PET-coated PVdC	—	5.5–7.3 10^{-3}	—	—

[Source: Karel and Heiss (89-77)]; *Packaging Handbook* (234); Bakker (188)]

With such data it is relatively easy to calculate the permeability B_{tot} of a multi-layer system. Let us consider a laminate composed of three layers with given thickness x_1 (cf. Figure 1.15). The total flux (F_{tot}) for a surface S is determined by the material with the lowest permeability constant (B).

$F_{tot} = F_1 = F_2 = F_3$
$S_{tot} = S_1 = S_2 = S_3$
$x_{tot} = x_1 + x_2 + x_3$

The general equation results in

$$F_1 = \frac{B_1.S_1}{x_1}(p_1-p_2) \quad \text{or} \quad (p_1-p_2) = \frac{F_1.x_1}{B_1.S_1} = \frac{F_{tot}.x_1}{S_{tot}.B_1}$$

$$F_2 = \frac{B_2.S_2}{x_2}(p_2-p_3) \quad \text{or} \quad (p_2-p_3) = \frac{F_2.x_2}{B_2.S_2} = \frac{F_{tot}.x_2}{S_{tot}.B_2}$$

$$F_3 = \frac{B_3.S_3}{x_3}(p_3-p_4) \quad \text{or} \quad (p_3-p_4) = \frac{F_3.x_3}{B_3.S_3} = \frac{F_{tot}.x_3}{S_{tot}.B_3}$$

$$F_{tot} = \frac{B_{tot}S_{tot}}{(x_1+x_2+x_3)}(p_1-p_4) \quad \text{or} \quad (p_1-p_4) = \frac{F_{tot}.(x_1+x_2+x_3)}{B_{tot}.S_{tot}}$$

Hence, by addition

$$p_1-p_4 = \frac{F_{tot}}{S_{tot}}\left(\frac{x_1}{B_1} + \frac{x_2}{B_2} + \frac{x_3}{B_3}\right)$$

$$p_1-p_4 = \frac{F_{tot}}{S_{tot}}\left(\frac{x_1 + x_2 + x_3}{B_{tot}}\right)$$

$$B_{tot} = \frac{x_1 + x_2 + x_3}{\left(\frac{x_1}{B_1} + \frac{x_2}{B_2} + \frac{x_3}{B_3}\right)}$$

If the moisture content of the film influences the permeability to gases, these equations must be used cautiously. Evidence of this is given by the permeability to carbon dioxide (CO_2) of a laminate of cellulose and polyethylene (LDPE), which depends on whether its cellulose side or its PE side faces the high humidity (cf. Table 1.18).

Using the data (e), (f), (g) in Table 1.18, the permeability can be approximated as follows:

$$B_{tot} = \frac{2x}{x\left(\frac{1}{B_1} + \frac{1}{B_2}\right)} = \frac{2.B_1 B_2}{B_1 + B_2}$$

Hence: $B_{tot} = (2.10.50\,000) \div (50\,000 + 10) = 20$

Although the value 3500 (h) should not be calculated by approximation, but worked out from the data (b).

For films connected in parallel the permeability of the composite can be calculated as follows (see Figure 1.15):

$$F_{tot} = F_1 + F_2 + F_3$$

$$F_{tot} = \left(\frac{B_1S_1 + B_2S_2 + B_3S_3}{x}\right)(p_1 - p_2)$$

$$F_{tot} = \left(\frac{B_{tot}S_{tot}}{x}\right)(p_1 - p_2)$$

Hence by equivalence

$$B_{tot} = \frac{B_1S_1 + B_2S_2 + B_3S_3}{S_{tot}}$$

If there are some pinholes in the film, the exchange of gases and vapours will not result from an activated diffusion but will follow a quite different law. Here, the pinholes are considered as small canals (capillaries); we can then work out the penetration of gases by Poiseuille's law. Another point to mention is that the permeability does not increase linearly with temperature but as an exponential function, the form of which is:

$P = P_0e^{-k/T}$.

Table 1.17 Permeability constant B, relative to water vapour of various packaging materials (37.8°C – 95% RH)

Material	$\frac{g.mil^*}{day.100\ inch^2}$	$\frac{g.100\ \mu m}{day\ m^2}$
Cellophane	10–100	40–394
Cellulose film + nitrocellulose coating	0.2–2.0	0.8–8
Cellulose film + saran coating	0.1–0.5	0.4–2.0
LDPE	0.8–1.5	3.2–6.0
HDPE	0.3–0.5	1.2–2.0
Saran	0.1–0.5	0.4–4.0
PVC	0.5–8.0	2.0–32.0
Al-foil 9 μm	0.1–1.0	0.4–4.0
Al-foil 35 μm	0.1	0.4
Paper-Al-PE laminate	0.1	0.4
Waxed paper	0.2–15.0	0.8
Polyester film	0.8–1.5	3.2–6.0
PP (oriented)	0.25	1
PP (non-oriented)	0.6–0.9	2.5–3.5
PS	6–10	25–40

[Source: Karel (89); Paine (220)]

* US Convention for 0.001 inch

Preservation of aromas for most foodstuffs like the preservation of perfumes for cosmetics, is an important quality criterion. The loss through the packaging cannot be calculated because the analytical methods to trace very small quantities of such aromas in the air are often inefficient or too expensive. Permeability to such products through packaging films is thus empirically determined by means of sensorial observation. Finally plastic packaging surfaces not only transmit gaseous components but can also diffuse liquid ones – the sticky or oily feeling coming from plastic bottles in which oils have been stored for a long period may be the consequence of this phenomenon! Such a loss is not important for foods but might have some importance for pharmaceutical preparations containing highly diluted 'active' products.

Table 1.18 Influence of moisture on the permeability to CO_2 of cellulose film (C), polyethylene (PE) and PE/C laminates

Material	Side facing the high humidity	Relative humidity (%)	Permeability $cm^3.x_1/m^2.day.bar$ $x_1=x_2=25$ µm	
Cellophane	–	0	10	(a)
Cellophane	*	100	3000	(b)
Polyethylene	–	0	50 000	(c)
Polyethylene	*	100	47 000	(d)
C + PE	PE	0	18	(e)
C + PE	PE	100	18	(f)
C + PE	C	0	18	(g)
C + PE	C	100	3500	(h)

[Source: Karel (89)]

There are many publications (89, 10, 12, 13, 35), drawing attention to the potential danger of plastic packaging for liquid medicines: on the one hand, an active component can be lost by sorption by the contacting film surface, followed by diffusion through the body of the film, which could be prejudicial in the case of weak concentrations. On the other hand, it might happen (as we will investigate later on with the migration concept) that components from the packaging migrate into the liquid medium so that the properties of the medicine differ from the original.

b) Selling and informative function of packaging

Emotional and pertinent elements

Increasingly, since the development of supermarkets and self-service selling, more and more goods are pre-packaged. Once the self-service system became established, an additional special function was assigned to packaging to replace that of the salesman – packaging must not only give information on the packaged product, it must also promote it and provide an incentive to buy. Thus, packaging has an emotional function giving the consumer the impression that an item is the 'best choice'. By means of its shape, colour, illustrations and printed information it must be attractive and informative. On average a consumer spends 20 to 30 minutes in a supermarket to buy 21 items (44) and therefore products must be easy to recognize. In this respect, shape and colour are essential. Most packages for drinks are cylindrical; but liquid pharmaceuticals for external use are presented in polyhedral (e.g. octagonal) bottles. Of course, custom and promotion play a role; however the shape of the packaging remains an excellent recognition signal and this is surely valid for blind people. Colour is more psychological – blue tints are equated to neatness and freshness, green to peace, yellow to strength and liveliness, red and orange to warmth, pastel tints give a soft, feminine feeling, whilst high primary tints equate to masculinity. Those associations depend on the culture of the population. Sacharow and Griffin (146) point out that in Eastern countries the symbolism of colours has an important significance in everyday life and that this must be recognized in packaging design. For the European purple is evocative of royal dignity, for Chinamen it means sorrow (see Table 1.19). Consequently it appears that, in the case of export goods, a pertinent preliminary survey may supply useful information.

The consumer judges most fresh foodstuffs on their appearance at first glance; the preservation and/or the display of natural colours are thus important and will have a deep influence on the saleability. Hence, a light transparent material is recommended for the packaging of fresh produce, although it would be preferable to use a darker packaging to limit the influence of light. However, it is common practice to prefer transparent packaging for light-sensitive products such as meat or milk. For meat this paradoxical solution is acceptable

Table 1.19 Symbolic meanings of some colours and pictures in packaging for the Far Eastern markets

Country	Colour	Meaning	Picture	Meaning
China	White	Sadness (a)	Tigers, lions and dragons	Strength (b)
	Blue	not good unpopular (a)		Strength (b)
India	Green and Orange	Good (b)	Cows	Holy for Hindus (a)
Japan	Gold, silver, purple,	Luxurious and high quality (b)	Cherry tree blossom	Beauty (b)
	Black	Only for printing (bright tints advised)	Chrysanthemum	Royal (a)
Malaya	Yellow	Royal (a)	Cows	Holy for Hindus (a)
	Green	Islamic Religion(a)	—	—
	Gold	Long life (b)	Pigs	Dirty and impure for Muslims (a)
Taiwan	Black	Mourning (a)	Elephants	Strength (b)
Thailand	Black	Mourning (a)	Elephants	National emblem (a)

(a) inadvisable; (b) advisable
[Source: Sacharow and Griffin (146)]

because the light penetration is limited to a small superficial layer, however the packaging must be impermeable to oxygen – *a priori* a certain inconsistency seems to exist in that case also. The purple pigment myoglobin giving its aspect to fresh cut meat, under the influence of atmospheric oxygen, turns to the dark red oxymyoglobin we know pretty well and often prefer. In an atmosphere containing oxygen, the oxymyoglobin will, however, form the undesirable brown metmyoglobin. Fresh meat is now preferably packaged in MAP conditions using high oxygen to promote red colour rather than using colour stabilizers. On the contrary, milk has no protection against the influence of light; in transparent packaging it loses a large amount of its vitamins within a short period (two to four days) (see also page 35). Packages impermeable to light are therefore a major requirement if we wish to limit the vitamin drop. Nowadays this can be achieved by using an Al-foil laminated carton. This solution raises other problems pertaining to the feasibility of waste recycling and to the migration concepts that we will investigate later. Dark-coloured glass offers good protection against light, is more convenient for the present waste and recycling processes, does not involve migration problems, but there is a strong chance that it would be less acceptable to the consumer.

Well-conceived illustrations can have a favourable influence on product saleability: e.g. a picture of snow flakes is meaningful in cold countries but of little importance in the tropics as it does not fire the imagination of the people living there. Sacharow (146) mentions that in Japan the sale of US tunny fish in cans was not successful because the fish was pictured with its nose looking downwards as if it were diving into the waves. This picture made the Japanese consumer think the fish was dead before capture. After modification of the picture, sales returned to the normal level.

In some particular cases, motifs are drawn on packaging for safety reasons, e.g. to point out the toxicity, the flammability,

Figure 1.16 'Mister Yuk' sticker as dissuasive.
This sticker painted with phosphorescent colour has been designed by the Southern Poison Center to dissuade children who cannot read the indications for use

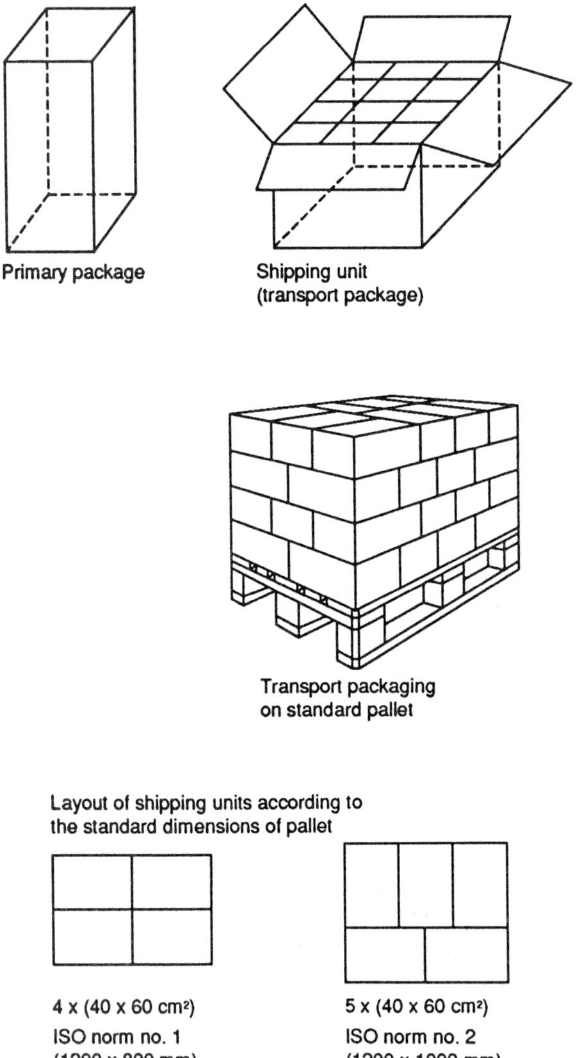

Figure 1.17 Schematic illustration of modular unit loads

the danger resulting from inhalation or from contact with the skin. The Southern Poison Center has designed a phosphorescent sticker to dissuade children from using a product (cf. Figure 1.16, 'Mister YUK-sticker'). By nature, children are inquisitive and in most cases they will taste, sniff and try the products they see for the first time: therefore it is wise to put these stickers on most toxic non-edible household products such as cleaning materials and polishes, detergents and washing powders, acids, alkalis, organic solvents, etc. It is obvious that the appearance and shape of packaging may be misleading (e.g. when it gives the impression of a large size). On the other hand, some small products are deliberately placed in bigger packages both to provide space for the necessary information and to make shoplifting more difficult.

Shape is not just an emotive way of identifying the product to induce time saving for the consumer. As explained earlier, shape can also save space, it can ensure efficient stacking and distribution, and consequently it results in savings in warehousing, transport and handling costs. The dimensions of most rectangular boxes designed nowadays are based on the 60 × 40 mm standard module. They are adapted to pallets the base of which is either 1200 × 800 mm (ISO No1) or 1200 × 1000 mm (ISO No2).

Unit loads

The loading of a pallet with primary packages, collective and shipping units to form a unit load is illustrated in Figure 1.17. Many different pallet patterns are shown in Figure 1.18.

Pallets are mainly made of wood; the unit load is secured with plastic strapping, stretch or shrink film, etc. Because of the savings created by modular packages, unit loads are more and more used even to the final retail outlet. In supermarkets, goods will thus be displayed in their unit loads on pallets. This way, the labour cost for unloading and displaying can be saved. Of course, alternatives for the indication of the individual price of the package have to be found; this will be discussed later when we deal with the control function of packaging.

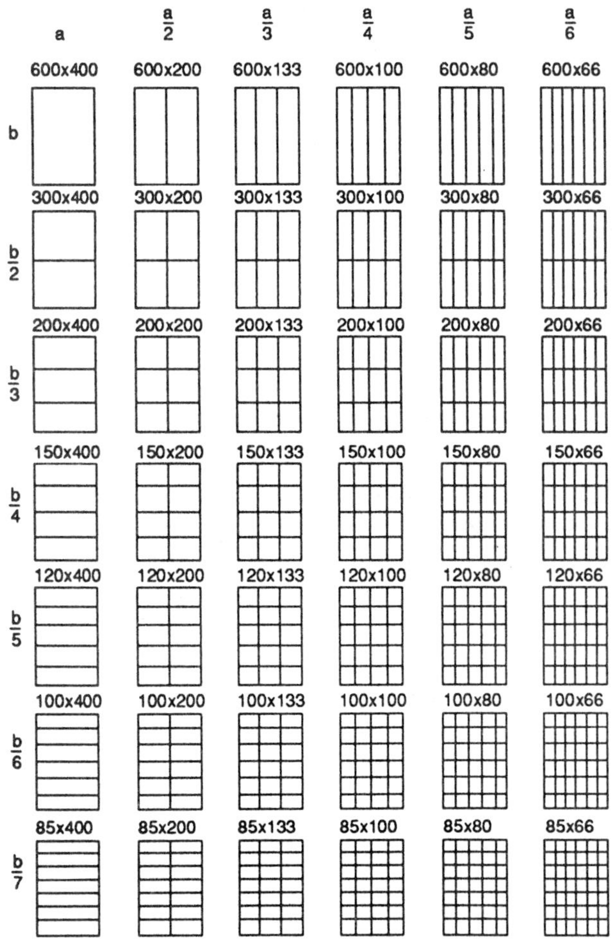

Figure 1.18 Packaging dimensions adapted to unit loads (60 × 40 module) [Source: Fink (53)]

Labelling

All this goes to show that nowadays the consumer must be given more details than ever before on the packaged product via the information printed on the packaging or the label. Since the publication of the Royal Decree (RD) dated 17 April 1980, modified by RD of 17 July 1980, and the RD dated 2 October 1980 (Belgian State Journal dated 11 October 1980, modified by RD dated 28 March 1983, Belgian State Journal dated 10 May 1983) all of which relate respectively to 'publicity for foodstuffs' and 'labelling of prepackaged foodstuffs', description and promotion have been defined in accordance with the EC's Directives 79/112 of 1978. The decree was to come into force on 1 January 1982, but was replaced by the RD of 13 November 1986 (published 2 December 1986), and later amended by the RD 9 February 1990 and the RD 4 September 1990 concerning the announcement of the party to which the sold foodstuff belongs. The decree is, with a few exceptions, applicable to all prepackaged foodstuffs.

Hence labels on food packaging must include the following:

1. Trade name of the product.
2. List of ingredients and additives (identification according to E-numbers.)
3. Sell-by or use-by date.
4. Special requirements for preservation and use.
5. Name and address of either the manufacturer, or the packager, or the retailer situated in one of the EC countries.
6. Directions for use, if necessary.
7. Place of origin of the foodstuff if its omission could be misleading.
8. Net quantity.
9. The alcohol content for drinks when it exceeds 1.2% in volume.
10. The indication of a number or party to which the foodstuff belongs.

Consequently, no description and/or promotion on the packaging related to health impact is allowed. (cf. RD of 8 January 1991, published 21 February 1992). It concerns in principle the indication of the caloric value of the product (kJ/100g), some nutrients (proteins, lucides and lipids), fibres and minerals. In some cases the 'advised daily consumption' should be

Figure 1.19 Symbol to be affixed on articles and materials to be in contact with foodstuffs (RD of 30.6.1981, EC-directive 80/590 of 9.6.1980)

mentioned as well. Since the RD of 10 July 1990 prices must also appear on the package. According to the EC's directive No. 80/590 dated 9 June 1980, the article designed to be in contact with foodstuffs must either state 'designed for foodstuffs', or carry the distinctive symbol shown in Figure 1.19. Finally we must mention that indications relative to volume, weight or quantity of prepackaged goods are also regulated and subject to the provisions of the laws dated 14 July 1981 (Trade Practices – Department of Trade) and 24 January 1977 (Law on Foodstuffs – Public Health), the RD dated 28 December 1979 and the ministerial decrees dated 12 September 1980 and 22 April 1982 (Department of Trade). These deal mainly with the compulsory statement of net quantities expressed in standard units and of the firm that filled the package. For prepackaged goods, the RD dated 28 December 1979 allows the EC sign '℮' placed near the statement of package weight and volume. When this sign appears on the packaging, the filler or the importer is responsible for the quantity indicated: they must guarantee a minimum variation in the content. This ranges from ±9% of the nominal quantity for packages between 5 and 50 gram (or millilitre) to 1.5% for packages between 1 and 10 kg (or litres). Hence, packaged goods with a stated weight of 300 g must never contain less than 291 g, those with a stated weight of 500 g must average 500 g and must never contain less than 485 g.

It would serve little purpose to describe here all symbols, signs and abbreviations used for labels. There are some legal requirements for almost every foodstuff – for instance, about the size of the printed information. For each product, some special abbreviations exist to define the quality: well-known are Cat I, II or III for beers; perhaps we are not aware that the density classes I, II and III refer to the strength of the sugar solution used to brew the beer. For eggs, a combination of capital letters, digits and signs is used to define a particular quality: the letters A, B and C refer respectively to freshness and allowable conservation period, the digits (1 to 7 inclusive.) indicate the weight group, 1 for eggs weighing more than 70 g,

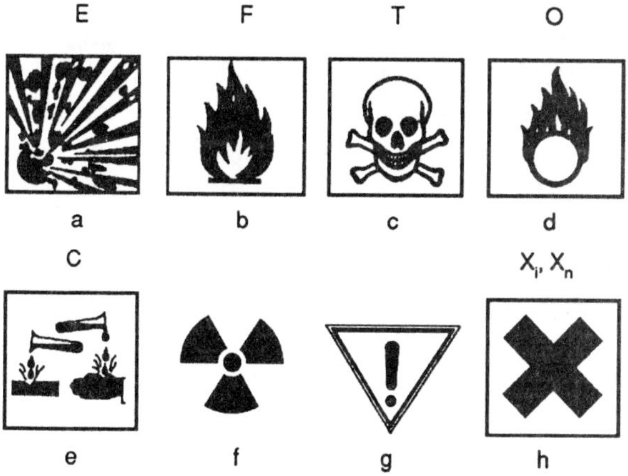

a: explosive
b: flammable
c: toxic
d: oxidizing
e: corrosive
f: radioactive
g: dangerous
h: noxious (X_n) or irritant (X_i)

The indications E, F, T, O C X_i and X_n are abbreviations of Explosive, Flammable, Toxic, Oxidization, Corrosive, Irritation and Noxious

Figure 1.20 Some symbols to point out dangerous goods (RD 13.2.1981)

2 for the group between 65–70 g, 3 for the group between 60–65 g, etc.; 7 is used for eggs weighing less than 45 g; both indications must appear in either a circle, a vertical rhomb or an equilateral triangle referring to their processing: the circle indicates that the eggs have not been refrigerated below 8°C. Normally class A eggs should be marked with a circle. A circle for class B means 'not refrigerated or long conservation eggs' although they do not meet the requirements for class A. The equilateral triangle refers to 'refrigerated eggs' and the rhomb to long conservation obtained by another process than refrigeration.

According to the RD dated 13 February 1981, a wide range of symbols and/or drawings are now applicable to indicate the nature and the danger of dangerous goods (cf. Figure 1.20). In addition to those drawings, numbers combined with R or S are mentioned: the R-number refers to the sort of danger reactivity and/or toxicity of the product, the S-number to the safety precautions to be taken.

At least the preservation of food by ionizing radiation (αrays, β–rays) must be made identifiable with an internationally recognized vignette (see Figure 1.21) printed upon the package. The commercial importance of food submitted to irradiation has for some time been regarded with criticism and in fact neglected for many years: the use of low dosages to a maximum of 10 kGy does not induce any form of radio activity in the irradiated products; also, only officially recognized agencies are allowed to perform the radiations for food prevention purposes. In the USA, irradiation is unfortunately classified as a food additive and regulated by the FDA. In most Western European countries (e.g. Belgium, the Netherlands, France), Japan and Canada, this treatment is being accepted and used for well defined, mostly solid, products (e.g. mushrooms, spices, some animal food) under carefully controlled conditions. In the UK, ionizing radiation and the import of products being so treated is prohibited, but since the report from the UK Advisory Committee on Irradiated and Novel Foods (No. 86) general acceptance can be foreseen at the opening of the single market and also with the statement of the WHO and the UN

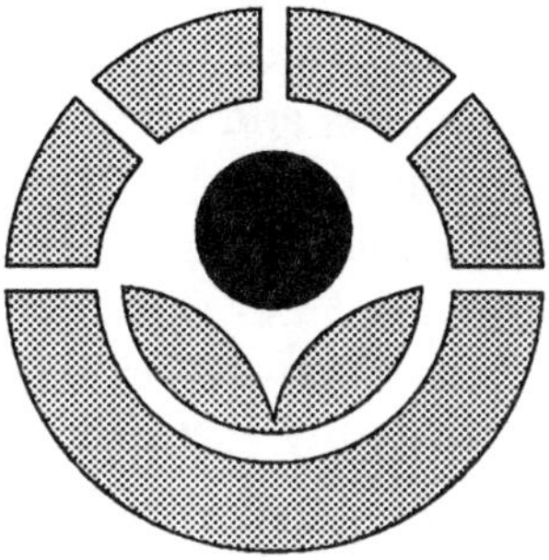

Figure 1.21 The internationally recognized vignette for irradiated foods

Food and Agriculture Organization's Joint Expert Committee mentioning that 'irradiation up to an average dose of 10 kGy presents no toxicological hazard and introduces no special nutritional or microbiological problems'.

A list of the major European directives relative to packaging is given in the Appendix.

c) Control function of packaging

Since the late 1970s a new function has been added to the already large range of packaging functions: this relates to the automation of data processing, mainly to improve the efficiency of company management. With computers, various calculations related to goods may be performed faster, but definition of the data and its input into the computer base remains a time-consuming task: selected data concerning the number of products received, the total number of parts in stock, ordered, displayed and/or sold must still be determined manually before automatic processing. In self-service shops, some still have a price label on every item (printed, self-adhesive etc.) that the cashier must register afterwards with a special product identification code. Considering the expanding trend of supermarkets, the larger the variety of marketable goods, the faster the flow of products induced by the systematic reduction of warehousing areas (40% in 1960, only 10% in 1970) (76), and the steadily increasing offer of exotic products, it has become necessary on the one hand to devise an international coding system to identify the articles, and on the other hand to invent an automatic system able to read it. First, national coding systems were developed such as the UPC (Uniform Product Code) in the USA and the PLU (Product Look Up-code) in Europe. In 1978, the EAN system (European Article Numbering) was successfully introduced. In 1982 it was already printed on their packages by more than 13 900 manufacturers in 18 countries and in 1992 it evolved to 200 000 and 53 respectively (20). The code consists of a 13-digit number, and the digits are split in four groups (cf. Figure 1.22(a)); the two first digits (f1, f2) identify the country of origin: 54 for Belgium, 87 for the Netherlands. (See appendix for other countries). The next group of five digits (x_1–x_5) is the manufacturer's registration number. This number is allocated in Belgium by the ICODIF (Instituut voor de COdering van

t_1^* t_2 x x	x_1 x_2 x_3 x_4 x_5 x x x x x	p_1 p_2 p_3 p_4 p_5 x x x x x	k t
Country number *2 digits*	Manufacturer's number given by ICOF-Belgium *5 digits*	Product number given by the manufacturer *5 digits*	Check number *1 digit*

t_2^* is not included in the barcode

Example a: codes of two different products (nos 40 and 41) of the same manufacturer (no 281) settled in Belgium (no 54)

5 410281 000401

5 410281 000418

Example b: a code consisting of eight digits comprising four for the manfacturer's registration number (no 2001) and three to identify the product (no 258). The last digit has a check function

2001 2588

Figure 1.22 The European Article Numbering, abbreviated to EAN-code, with the relevant bar code legible omnidirectionally for optical detection and electronic data processing: a) examples of a 13-digit code; b) example of an 8-digit code

DIstributeurs en Fabrikanten), in the Netherlands by the UAC (Stichting Uniforme Artikel Codering). The third group of five digits (p_1–p_5) identifies the product; this number is allocated by the product manufacturer. The final digit (k) is a check number derived from the previous ones. Each marketable product is thus uniquely identified by its number; multi-packages (e.g. three chocolate bars in another bag) are given a new code number differing from that of the unit article. Codes with 14 digits are now produced for special purposes (e.g. identification of transport packaging). In accordance with the EAN system, some self-service businesses or supermarkets are allowed to use a special code for products they sell under their own brand name. This code consists of eight digits: four digits for the registration number of the company having manufactured, processed or packaged the goods and four digits for the product and control number (see Figure 1.22 (b)).

The code number is then translated into a bar code for automatic optical reading; each digit is represented by a combination of two black lines and two white spaces. By varying the thickness of two adjacent lines and the distance between them, it is possible to represent the 10 different digits (0–9). Two narrow lines, respectively the start and the separation signs, indicate the beginning and the end of the code and also separate the company registration number from the product number. As soon as a package provided with the code is moved correctly over a sensitive photocell, in such sense and direction (almost omni-directional) that all parts of the code can be detected, the light pulses induced by the black lines and the spaces are converted into electrical pulses, a language understandable by a computer. Instead of 'reading windows', 'reading pencils' can be used; their flexibility is higher, especially when handling voluminous or heavy packages. The signal received by the cash register for a product is automatically compared by the computer to data, price and description of that product, in its memory. These data are then automatically printed on the bill. It is thus no longer necessary to price every article. The immediate advantage for the consumer is the fact that goods can flow very quickly at the check-out resulting in a shorter waiting time. However, the advantages of the system for control and management consist mainly in further opportunities: each article registered is an article taken away from stock; the central computer collects all data relative to purchasing and selling and can give constantly the state of the stock and the level of inputs and outputs, data which can afterwards be processed statistically. It is obvious that various consequences and influences on the turnover, for instance those resulting from special offers, can be figured or rectified rapidly. Reading pencils as mentioned earlier, can also be connected to computers increasing the mobility for fast checks, e.g. for packages and parcels in lorries, at the custom house, etc.

The article coding system has disadvantages also: consumers no longer see the price of a product on its package; of course, the price of the article purchased is printed on the bill but generally the check is only made when the purchaser reaches home. A comparison between the price charged and the price indicated on the shop shelves is now more difficult. Of course, a unique code is inapplicable to packages the weight of which vary as for meat, fruits and vegetables; nevertheless, balances have already been constructed to perform an automatic data printout following weight and kind of product on labels to be stuck upon the closed wrapping holding these goods.

For some purposes, digits and signs are printed on the label that are quite incomprehensible to the purchaser: they relate to the manufacturing date, sometimes referring to a batch number. This code is not important for the distribution sector, it has no link with legal quality requirements or provisions, but will be recognized only by the manufacturer. Such codes are, however, quite important at the production level, especially to cope with claims. A sequence of small notches in the edges of labels for beer bottles is a typical example of a non-numerical code.

Finally we must mention that colour codes exist also; these consist of a line of dots or vertical stripes and are not combined with digits; until now they have mostly been printed on pharmaceutical packaging. They also mainly relate to manufacturing data.

d) Packaging as an identification element for waste treatment

We have already explained that polymers used in packaging can be 'made to measure'. The packager's predetermined wishes as regards to tensile strength, light penetration and impermeability to gases can be met by an adequate choice of chemical reactants, by modification of the reaction conditions (pressure, temperature and catalysts) and finally by a well-ordered combination of films and foils in laminates, plastic coatings etc.). Due to the large variety of these combinations, the recycling of mixed packaging scrap and/or household refuse which includes approximately 7% plastic packaging, is not easy, especially if we are concerned with 'material' recovery.

Recycling of plastics consisting of a mixture of various individual polymers, etc. is a difficult and expensive task; it becomes virtually insurmountable when the recovery process must treat soiled plastic packaging after it has been separated from household refuse. For these reasons, incineration to recover energy as heat seems to be the best alternative. When waste consists mainly of one plastic group, even soiled, more efficient recovery should be possible.

The latest relevant ideas involve plastic packaging being 'marked' according to their group, for identification and/or automatic sorting in order to be able to use the most favourable recovery method. Some PS-cups used for yoghurts and creams are already marked with three arrows arranged in a triangle, in the middle of which appears the material symbol PS. A colour code consisting of a colour strip on plastic packaging could be another possibility. Such marks have already been used on PET bottles for soft drinks: they were in fact placed on the coloured base, but for most the material of body and base were different! Since 1987 PET bottles have been designed to stand on their own and some have become reimbursable (e.g. in Germany and the Netherlands) when returned to the shop. The most recent development concerns the identification of so-called 'environment-friendly' packaging. The adjudgment of such an 'ecolabel' is based upon the life-cycle analysis (LCA) of the packaging material. Ecolabels have been produced in

Germany (the Blue Angel), Canada, Japan (Ecomerk), Australia (green print), the US (green stamp, green cross). Since 1992 an Ecolabel symbol has been adopted for the European products as well (see Figure 1.23). The main purpose is to encourage manufacturers and consumers to use and choose materials for a certain functionality with a recognized low impact on the environment. In Chapter VIII, which deals with waste management, we shall discuss more about the aspects of LCA.

Figure 1.23 Some examples of ecolabels
(a) Canada; (b) Germany (the so-called 'Blue Angel'); c) Europe: European ecolabel accepted March 1992 by the European Council to designate environment-friendly products and their packaging

II
Environment

1 DEFINITION

The environment includes all those elements of this planet Earth that create an astronomical body supporting a large range of living species. It includes not only the layer of air around the planet but also the entire atmosphere, the composition of which is fundamental for life on Earth. The ozone layer (present at 30 km above the earth's surface) (161) filters most of the UV light emitted by the sun in our direction. Major disturbance and/or damage to the ozone layer by Freons (components of the propulsive aerosol gases that, according to Rowland and Molina's theory, induce through a sequence of reactions a progressive reduction of that layer) will modify the structure of many living species. Also the electrical charge of the ionosphere influences the quality of the space where we live; major modifications to it can indirectly affect our health.

This definition implies also that all world resources crucial for life in any economic system, such as minerals, water, air, energy (coal, crude oil, etc.) must remain available. The 'environment' concept could also be approached from another angle related to the relationship between human beings and their surroundings. The 'surroundings' may be urban or rural and involve another relationship with nature. It could also be concerned with the society in which we live. Here we must take into account the sociological, economic and political influences and disparities. To describe the potential connections with packaging, we will consider the concept of the environment in every sense of the word. We have already examined the social implications and the economic importance of packaging, so no great attention will be given to those aspects in the following discussion.

2 PRINCIPLES OF NATURE'S MATERIAL EVOLUTION AND RECYCLING

At the 'top' of all living organisms is homo sapiens, differing from all other species by his consciousness alone. At the bottom of the scale is the vegetable kingdom with its distinctive ability through chlorophyl to connect water, carbon dioxide and solar energy into solid substances: glucose, starch, cellulose, etc. At the same time oxygen, so important for life, is produced from the water and carbon dioxide (see Figure 2.1). All other living species fall between these two extremes. It is also important to be aware that nature forms, throughout the succession of its manifestations, a cyclic food chain and that reciprocal dependence does exist. From a strictly materialistic point of view, the food chain has no beginning or end; every organism has its own life cycle always including birth, growth and death, during which substances (foodstuffs) are ingested for growth and energy supply. The wastes produced during life and after death are broken down by enzymes and then converted by bacteria to be again assimilable by plants. Additionally, parts of most organisms can serve directly as foodstuffs for other organisms.

During the respiration process, the oxygen produced by plants is used to create water and carbon dioxide again (the heat generated by this biological oxidation process may be considered as recovering solar energy). Subsequently the level of water pollution by biologically degradable wastes may be expressed by the amount of oxygen needed by the microorganisms to break down and treat them (BOD coefficient: biological oxygen demand – cf. also the respiratory quotient RQ for the treatment of household refuse by composting). Throughout nature we have a succession of all those incomplete life cycles, during which simple molecules (e.g. water, carbon dioxide, nitrogen and phosphorus compounds), used as

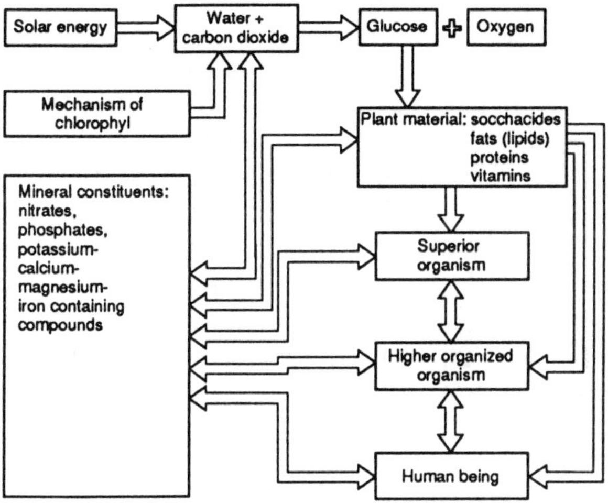

Figure 2.1 Diagram of material cycles through living organisms

material by increasingly more complex living species, can be transformed into the most sophisticated molecules and then be metamorphosed again; a cyclic evolution sustained by solar energy. With these biological breakdown reactions (biodegradation), nature gives us a good example of recycling material.

Although in this cycle of substances (foodstuffs) many chemical reactions are induced by the living organisms, the entity (of each organism or the complete cycle) differs clearly by its structure from a pure chemical reaction, the evolution of which is characterized by a maximum increase of the entropy relative to the system in question. Physically this means that, whilst spontaneous and promoted chemical reactions always induce more chaos, conversely, in nature, the final result of such reactions corresponds to a minimum entropy variation. It seems likely that 'keeping order' requires energy. This is valid not only for the formation of every organism and for the biological consistency and cycle, but also for the structures built by people (e.g. buildings and towns) and for the organization of societies (organization of firms, communes, provinces, etc.). We may therefore assume that the evolution of packaging materials, for instance by the management of recycling, will tend to some natural order.

The environment can also be considered as a dynamic biological entity, the structure and equilibrium of which remains stable because nature with its 'waste treatment' system has created a 'material recycling' process; that mechanism is usually regarded as a 'decontamination process', starting as soon as 'contamination' occurs. For natural wastes and their processing, nature has a buffer capacity ensuring the neutralization of wastes caused by temporary variations in densities of population. For instance, an excess in the food supply of a pool that was originally in equilibrium, provokes the death of fish. A large amount of foodstuff will always raise the BOD coefficient of the system, increase the density of the bacteriological population in order to 'consume biologically' the oversupplied products. The necessary oxygen is present, however, in small quantities only because of its low solubility in water; hence there is less oxygen available for the fish, resulting in a decrease in their life expectation. After a while, the original equilibrium may be restored; the excess bacteria will then be dead and most of them will be lying as mud on the bottom where other anaerobic breakdown mechanisms will recommence. The oxygen content of the system enriched in minerals will rise again due to the growth of plants (algae, for instance). The eutrophication of rivers can also be considered as a natural decontamination process consequential to a large supply in phosphates: however, in that case, a new system with a new equilibrium will be created.

Similar phenomena occur in the soil also, on the ground and in the air: when equilibrium conditions exist, any contaminations are always neutralized by mechanisms inducing mostly a modification in the densities of some populations. Generally, the number of different organisms decreases while the population of a particular species increases. Natural equilibrium is mainly characterized by a large range of species, the total population of which fluctuates. As long as this fluctuation remains possible, the society in question has a buffer capacity to react against alien elements such as contaminants. If permanent damage occurs due to some influence and the natural repair mechanisms cannot work, we deal with pollution instead of limited contamination.

3 MATERIAL EVOLUTION IN INDUSTRIAL PACKAGING AND GOODS PROCESSING

Most articles for everyday use on the market nowadays are manufactured or handled on an industrial scale. Packages once generally made by hand are now also subject to industrial processing. Woven baskets etc. are quite rare nowadays. Works of art, jewellery or antique furniture can hardly be considered as industrial products although all should be protected by adequate packaging during transport. Farm produce (potatoes, vegetables and fruits) is also handled mechanically and packaged before being displayed as 'fresh' produce in the supermarkets. Marketable goods consist in the combination of products with their packaging, hence almost all have been subjected, either completely or partly, to an industrial processing.

To manufacture and/or handle such products and their packaging, mechanical and/or electrical power is needed as well as raw materials. Energy is also consumed in the carriage of the raw materials to either the industrial plant or to the distribution centre. The generation of mechanical power by means of combustion motors (stationary or movable) and of electric power, either in a classical power plant from the heat generated by the combustion of coal, gases or oil products, or in a nuclear power plant from the heat of nuclear fission, causes damage to the environment. Every industrial activity, including the production of packaging materials, has to cope with some waste and pollution problems. According to directives and/or decrees relative to the protection of public sewage and surface waters, the effluents of some industrial activities must meet the requirements of specific standards. They concern mainly the temperature, the acidity, the BOD and COD values and the percentages of floating and sedimentary particles.

For the emission of gases (e.g. sulphur dioxide and volatile components), there are applicable directives for Belgium; directives also exist for tipping and recycling of solid wastes. These will be examined in relation to packaging materials later. Packaging materials and/or forms judged defective during the production process are mostly recycled on the spot. Such material is considered as internal and/or new waste (see page 179) and does not affect the environment. Other sources of pollution include the noise generated inside and outside the plants (including the noise nuisance provoked by traffic), thermal pollution, light pollution and ionizing radiations emitted by radioactive products (in nuclear power plants, in chemical industries which purify and recover radioactive (waste) products, and in centres where packaged goods and packages are treated for sterilization purposes, etc.). These phenomena are not linked with the material but their psychological and physical effects can be harmful so that they may be considered as pollutants.

After being purchased from the retailers, products are separated from their packaging which can then become litter (waste dropped at an inappropriate place) or become part of the household refuse. Unlike industrial solid wastes, packaging wastes are scattered, stained, sometimes added to other packaging materials in heterogenous packaging refuse or mixed with other domestic refuse. For each material, the feasibility of recycling will be investigated in the following pages. At the moment, selective collection and gathering in special containers (bottle-banks for instance), sorting after collection and/or treatment of the collected wastes are being organized. The different methods of treating household refuse, whether or not they are linked to sorting mechanisms, are examined in Chapter V.

Collected solid household waste, including discarded packaging, makes a very specific material, which must be stabilized because of its biological activity. Industrial refuse, the refuse of agro-industry (e.g. solid wastes from farming) and effluents from water purification stations (mud) may, in particular cases, be treated together with household refuse. Such operations, viz. composting, combustion, compaction or pyrolysis (a kind of dry distillation), may also be considered as industrial activities

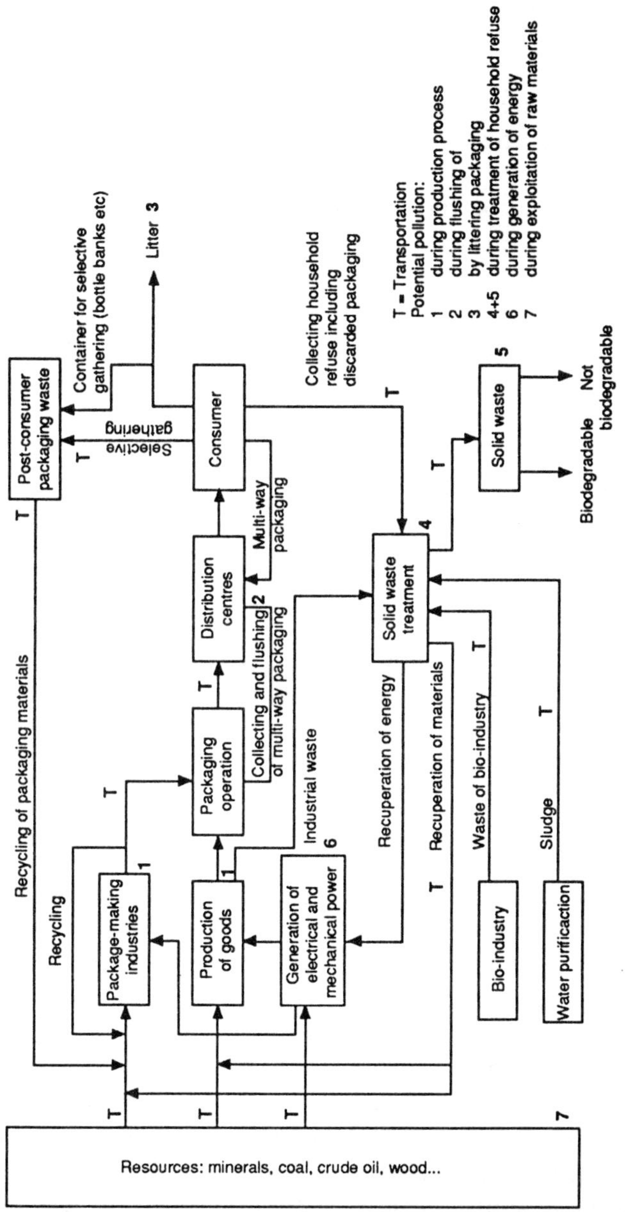

Figure 2.2 Pattern of the packaging life-cycle and of the potential interactions with the environment

requiring energy and featuring specific pollution problems. Moreover, every treatment, from the simplest process to the most sophisticated, involves the production of solid wastes that will be absorbed or fail to break down in the natural biological chain. Thus, the influence of packaging on the environment through industrial processes is a very diverse problem. That is the reason why we will use the packaging life cycle illustrated in Figure 2.2 to support our discussions. Returnable packaging – until now only glass has been really convenient for this application – is not usually part of the waste cycle as it is reusable. However, it is treated as waste when broken, and the washing of these packages before reuse also contributes to environmental pollution.

4 SUMMARY OF THE POTENTIAL INTERACTIONS BETWEEN PACKAGING AND ENVIRONMENT

The situation described above, points out that packaging can interfere with 'environment'. More details are given in the following seven points (the same numbering is used in Figure 2.2 for clarity in comparison):

1. *During raw materials production*: this is valid for packaging as well as for every industrial process.

2. *During cleaning of multi-trip packages*: (potential pollution of water by washing powders, for instance).

3. *As litter*: spoiling the appearance of the country, forests, beaches, parks, streets, rest areas along the motorways, etc. As most of its components, such as glass, metals and plastics, will not disappear through the natural cycle even in the long term by biological breakdown reactions, unlike paper, this kind of pollution can be very detrimental; there is indeed no technical solution to cope with this problem. Only education not to litter, and permanent information via the mass media may perhaps restrict this kind of pollution.

4. and 5. *During treatment of household refuse*: the remark in point 1 applies here also. However, we may assume that packaging can have a great influence on the choice of the

treatment method because its part in the solid household refuse amounts to approximately 30% of the total dry weight. Of course, the great variety and variability of selected materials partly induce the heterogeneous composition of the wastes. Besides, this composition may vary significantly with seasons, economic level, availability and/or price of raw materials and energy.

Taking all these factors into account, it is not easy to choose the best method of treatment; the lack of landfill sites and the fixed purpose to recycle and/or to recover materials or energy may be decisive. The presence of toxic products, traces of chemicals and sprays in the packaging, the presence of highly flammable products, of propellant gases in aerosols (Freons), of radioactive elements etc., imply the need for preventive regulations relative to waste treatment in order to avoid spoiling the environment.

It is clear that the solid waste from this 'industrial activity' depends on the composition of the collected domestic refuse and on the applied method; opportunities for its use and tipping must also be restricted.

6. *During generation of mechanical and electrical power and in transport*: the steps involving transport are indicated with 'T' on Figure 2.2. Imported raw materials are needed for the production of most types of packaging; of course, extraction and transport require mechanical and/or electrical power. The influence of the energy production on the environmental factors is detailed in Chapter IV. Packages will differ from one another according to the raw materials used and their energy consumption; some materials, e.g. paper, fibreboard and plastics, have a calorific content after production. Figure 2.2 shows also that packaging must be transported either empty, or full, or afterwards as waste: for actual packages, the transport cost is always proportional to the ratio of weight to effective volume.

Stacking density and effective surface area are lower for cylindrical glass or plastic packaging (bottles) than for rectangular ones; we have already pointed out that empty paper and board packaging can be stored, stacked and transported flat; so maximum effective surface area and saving in cost and energy are achieved. Transport is also a kind of burden for the environment: indirectly through the production of energy for electrical traction or directly by internal combustion engines (thermal and atmospheric pollution). The part packaging plays in the noise generated by transport is difficult to figure; noise nuisance exists also inside packaging plants due to handling (e.g. chinking of glass bottles).

Although specific details of 'packaging and energy' will be considered in Chapter IV, it is interesting to mention here the energy consumption pertaining to 500 ml bottles for soft drinks considered either as returnable packaging (8 fillings) or as throw-away packaging (see Table 2.1). Assume that a returnable bottle weighs 500 g; and the weight of a single-trip bottle, normally thinner and more fragile, is 330 g, consumption rates (76) show that most of the energy is used during manufacture and the complete packaging manipulations. Energy is also needed during recycling of the packages both for sorting and for transport to the bottle manufacturer. Analysing this case, it is clear that returnable bottles, not associated with material recycling after breakages, require far less energy than the use of throwaway packages. However, energy consumption is higher for recycling products than for landfilling. Therefore, raw material and energy costs may be determinant in the steady trend to produce more single-trip packagings.

7. *World resources and raw materials consumption in packaging production*: the lessening of the world resources in raw materials depends mainly on the world population and on its standard of living. The world population amounts to approximately 4 billion people and the average annual growth is about 2%. In 1966, world food production was approximately 33% higher than in 1951; during the same period the consumption of chemical fertilizers and pesticides rose by 146% and 300% respectively (27, 104). About 30% of the population live in towns (70% in the USA). There is no doubt that population growth and the continous trend to town living will involve a further increase in the use of packaging materials. A lack of raw

Table 2.1 Comparison table relative to the energy consumption (in 10^6 Joules) when using 500 ml glass bottles for soft drinks either as throw-away packaging or as multi-trip packaging (7× reusable)

Operation	Multi-trip packaging (8 fillings)	% total	Throw-away packaging (8 packages)	% total
Extraction of raw material	1.045	5.1	5.480	8.9
Transport of raw material	0.131	0.6	0.686	1.1
Production of bottles	8.163	40.3	42.860	69.9
Production of crown corks	2.041	10.0	2.041	3.3
Transport to filler	0.381	1.9	2.000	3.26
Filling etc.	6.435	31.7	6.435	10.5
Transport to distribution centre	1.983	9.8	1.303	2.10
Transport by the consumer	–	–	–	–
Collection of refuse	0.094	0.46	0.494	0.80
Total	20.273		61.299	
Energy for packaging sorting and recycling	1.163		6.100	
Total with recycling	21.436		67.399	

[Source: Hannon (76)]

materials for making packaging is probably unlikely (see other chapters); the situation is of course quite different for energy-producing resources such as coal (700.10^{12} ton), crude oil (107.10^{12} ton) and natural gas (90.10^{12} ton), which might run out within the next 50 years if the present consumption rate is maintained. Moreover, the energy consumption rate of any industrialized society always increases faster than its population.

5 ECOLOGICAL PARAMETERS

Because of the exploitation and the nature of raw materials, their transport and production, their energy consumption and waste treatment, the potential impact of packaging materials on the environment affects a large range of industrial activities. The influence of packaging on the 'environment' is generally difficult to determine and to separate from the influence of similar products generated by similar processes. For example, there is not much difference between the production of window glass and of glass hollow ware. Besides, many parallels may be discovered in the contamination of the environment by various industries, therefore it seems necessary to describe completely the main elements able to disturb the expectation of life in nature. Each time we study one of the main packaging materials, reference will be made to its potential impact upon the environment.

The main factors we try to present here in a logical order, will rarely occur alone: for instance, the environmental impact of the discharge of high BOD effluents into a river depends very much on the temperature – on the one hand, the toxicity of the products and the life rhythm of the micro-organisms taking part in the decomposition reactions will increase in direct proportion to temperature; on the other hand, a high BOD means an important need for oxygen. This is a need which increases with temperature, while lower quantities of oxygen are present in water due to poor solubility. A lack of oxygen in water will result in the death of higher organisms such as fish, inducing a further increase of the BOD-level. In a low-oxygen system an anaerobic fermentation begins that generates, for example, the malodorous and toxic di-hydrogensulphide gas (H_2S).

The potential implications of such a discharge are thus far more difficult to foresee than might appear at first sight from the composition of the waste water and the temperature. Particular incidences might even remain unnoticed or be considered negligible, especially when a 'migration' can take place and/or when the organisms coming from the polluting source can hive off. Plants exposed to toxic products in the air and in the soil cannot migrate for defence. However, we might consider a potential dilution in the air by the wind and in the soil by the rain.

We think it advisable that every request relating to the installation of an industrial plant should include a report giving the results and/or conclusions of an ecological survey relative to waste discharges. The Belgian Authorities have already enforced the departmental order dated 29 July 1982 relative to the requests to licensing for waste treatment; the application forms (Belgian State Journal 10.09.1982, p. 10480) include a heading concerning environmental management. The marketing of products dangerous for man and the environment is controlled by the RD dated 24 May 1982 (Belgian State Journal 02.07.1982): reference is made here to studies relative to carcinogenicity, toxicity, and mutagenicity.

a) Product toxicity

Definitions

The ecological toxicity we are interested in is relative to the harmful effects of products present in air, water, soil and food as well as other factors such as radiation and temperature. The main parameter to be taken into account is the dose, the amount of product in question per kg biomass (cf. Paracelsius's statement (1493–1541): 'every product is a poison, there is no product without any poisonous property: toxicity depends only on the dose').

The harmful effects of products are determined as follows:

1. Median lethal dose (LD 50): statistically determined quantity of a product which would provoke 50% death when ingested by a population of experimental animals.
2. Median effective dose (ED 50): quantity provoking a specific effect on 50% of the test population of experimental animals.
3. Acceptable daily intake (ADI): quantity which can be ingested daily without provoking any effect. In fact the potential risk is regarded as negligible in that case.

These quantities are expressed in mg/kg of body weight. The toxicity of poisonous gases and products present in air is prefer-

Table 2.2 Comparison table relative to LD 50 quantities of product in different countries

Country/class	Skin contact (mg/kg body weight)	Intake (mg/kg body weight/day)	Inhalation (mg/litre air)
EEC/very toxic	50	25	0.5
USA/very toxic	200	50	2 (200)
UN/class 1	40	5	0.5
EEC/toxic	400	200	0.5
USA/toxic	2000	5000	200 (solid)
UN/class 2	2200	50	10
EEC/dangerous	2000	2000	20
UN/class 3	1000	2000 (liquid) 500 (solid)	200

ably indicated by a concentration (mg/m³, ppm), generally valid for a specified period (8h/day; 5 days/week).

Classification of products according to their toxicity

Products can be classified according to their toxicity, depending on the method of ingestion (inhalation, oral intake, subcutaneous injection) and on the adopted system. Some specifications are given in Table 2.2 for comparison between different countries.

Factors influencing toxicity

The toxicity of products is influenced by both the physical state of the pollutant (i.e. solid, vapour, or liquid) and in the case of solids and liquids the particle or droplet size.

1. *Particle size of products*: The finer the particle size of the product, the greater the surface area and hence the toxic effect.

a) Solids: generally solid materials are not toxic; their toxicity depends on their solubility in water, or on the solubility of any derivatives created on their surface. Table 2.3 gives the toxicity

Table 2.3 Toxicity of some elements when present as ions in water

Element with chemical notation	LD 50 (mg/kg body weight)	ED 50 (mg/kg body weight)	Toxicity class
Sodium (Na^{+})	1 480 000	1 020 000	A
Calcium (Ca^{2+})	300 000	220 000	
Magnesium (Mg^{2+})	190 000	125 000	
Potassium (K^{+})	97 000	68 000	
Tin (Sn^{2+})	42 000	1500	
Iron (Fe^{3+})	5900	5200	
Arsenic (As^{3+})	2850	1400	B
Chromium (Cr^{3+})	2000	600	
Aluminium (Al^{3+})	1400	680	
Lead (Pb^{2+})	300	100	C
Zinc (Zn^{2+})	158	102	
Nickel (Ni^{2+})	130	95	
Copper (Cu^{2+})	44	35	D
Mercury (Hg^{+})	13	6.7	
Cadmium (Cd^{2+})	5	0.7	

A: not toxic; B: slightly toxic; C: rather toxic; D: very toxic
[Source: De Maeseneer (36)]

of some mineral components which may be present in effluents from industries (packaging production) or which may be liberated by extraction (leaching) from disposal sites. The table shows that sodium ions, present in kitchen salt, are not toxic. It is also clear that, even if the toxicity of a product is negligible, its use, application or discharge must be restricted to avoid any harmful effects. Quality standards of rivers and surface waters in general are summarized in Table 2.4. These figures are normally used to calculate critical volumes, the amount of pure water necessary to dilute the effluents from industrial processes in order to meet the quality standards.

Table 2.4 Quality standards of surface waters

Parameter (abbreviation)	Acceptable limit of presence mg/l
undissolved products	20
acids (pH-limits)	4.15 – 9.3
lead (Pb^{2+})	0.5
cadmium (Cd^{2+})	0.1
iron (Fe_3^{2+})	0.2
copper (Cu^{2+})	0.5
mercury (Hg^+)	0.01
ammonia (NH^3)	1
ammonium (NH_4^+)	
chloride (Cl^-)	1000
cyanide (CN^-)	0.1
fluoride (F^-)	10
sulfide (S^{2-})	0.1
dissolved organic compounds	10
BOD	20
oils and fats (saponification possible)	20
hydrocarbons (HC)	10
chlorine containing dissolvants and organic products (lipophylic)	0.1
phenols	0.05

[Source: Widmer (199)]

b) Vapours: vapours may be present in the air as free molecules. Some, such as vinylchloride (monomer of PVC), sulphur dioxide (generated by the combustion of sulphur compounds in fuels for transport and power generation), and carbon monoxide (from incomplete combustion of gasoline) are very toxic. Vapours, such as styrene vapour (styrene monomer) can also be dangerous. Consequently their permitted level in air is very low. Table 2.5 indicates some maximum concentrations of pollutants in air.

Using chemical methods, dangerous vapours can be extracted from the air; however, new combinations are created that are generally evacuated or discharged in aqueous solution.

Thus the pollution problem is shifted from the air to an effluent. Due to the heterogeneous composition of 'combustibles', present in household refuse, their incineration can, if the incinerator is not operating properly, generate dangerous gaseous products: the presence of chlorine compounds (e.g. NaCl, PVC, etc.) can give raise to chlorine and hydrochloric acid, products which can react in the gaseous phase with other components formed to create dangerous emissions (e.g. dioxins); besides soot (black smoke clouds), the

Table 2.5 Quality standard for air: permitted concentrations of solid particles and vapours in air according to the exposure period

Product description	Permitted maximum concentration
Solid particles (dusts and smokes)	75 μg/m³ average daily exposure 260 μg/m³ – max. concentration for 24h * 60 μg/m³ – ** 150 μg/m³ – average for 24h **
Sulphur dioxide	80 μg/m³ (0.02–0.03 ppm) 365 μg/m³ (0.14 ppm) max concentration for 24h exposure * 1300 μg/m³ (0.50 ppm) – max. concentration for 3h exposure * 60 μg/m³ – ** average daily exposure 150 μg/m³ – 24h exposure **
Carbon monoxide	10 mg/m³ (9 ppm) – for less than 8h exposure * 40 mg/m³ (36 ppm) – for up to 1h exposure *
Hydrocarbons	160 μg/m³ (0.24 ppm) – for up to 3h exposure *
Nitrogen oxides	100 μg/m³ (0.05 ppm) – average daily exposure
Vinylchloride	1.2 ppm (provided the total absorbed does not exceed 500 mg/kg body weight in one year)
Styrene	100 ppm

* This value may be exceeded once a year
** Data in accordance with RD dated 16 March 1983 (Belgian State Journal 29.04.1983)
μg/m³ microgram per m³ air
mg/m³ milligram per m³ air
ppm: parts per million, e.g. 1 ml vapour per cubic metre of air.
[Source: EPA (USA)]

incomplete combustion of most plastics can also generate toxic decomposition products. This explains why casualties can be high in case of fire in confined spaces where plastic materials have been used in furnishings and as wall and floor coverings for decoration. This demonstrates that incineration of household refuse and industrial wastes should be carried out carefully if air pollution is to be avoided. The ambient air quality standards for most exhaust gases encountered in industrial processing are listed in Table 2.6: these figures are normally used to calculate critical volumes, the amount of pure air needed to dilute emissions to meet the mentioned standards. These critical volumes are referred to when ecobalances of types of packaging are evaluated.

Table 2.6 Acceptable limits of gaseous compounds in the air

Parameter (abbreviation)	Acceptable limit (mg/m^3)
particulate matter (PM)	0.07
carbon monoxide (CO)	8
ammonia (NH_3)	0.5
nitrogen oxides (NO_x)	0.03
hydrochloric acid (HCl)	0.1*
sulphur dioxide (SO_2)	0.03
chlorine (Cl_2)	0.02*
fluoride (F_-)	0.1
hydrofluoric acid (HF)	0.05
mercaptans (- SH)	0.01*
dihydrogensulphide (H_2S)	0.015*
hydrocarbons (HC)	15*
aldehydes (cfr formaldehyde)	0.03
organic compounds (cfr ethylene-imines)	0.01*
lead (Pb)	0.001
mercury (Hg)	0.0007*
cadmium (Cd)	0.00001

* calculated from MAK – values
[Source: Widmer (199)]

c) Liquids: Van der Waals' forces (cohesive forces) maintain the molecules together. As far as the study of toxic liquids is concerned, they are often assumed to be organic. With a few exceptions (some alcohols) these have a rather poor solubility in water. Mists are formed by very small droplets in the air. Mist is generated by the condensation of vapours in the air (e.g. carbon disulfide during preparation of Cellophane, water vapour in the cooling towers of electrical power plants.) Small droplets can also give stable emulsions in the presence of detergents in water. Again the size of the droplets can have a significant influence on the toxicity of the product. Liquid organic pollutants related to packaging occur mainly in the preparation and processing of polymers and additives for plastics.

d) Smokes: very small solid particles the fineness of which enables them to remain suspended in the air for a long period (e.g. fine metallic oxides dissipated in the air during the iron refining process, fine particles of soot produced by incomplete combustion of plastic refuse). Such particles do not settle out easily by gravity but generally need physico-chemical processes.

e) Dusts: solid particles (rock, woodfibres, chemicals) present in the air as fine powder. Dust can be generated by the exploitation and/or the handling of ores, and chemicals (e.g. handling of limestone and soda in glass-making, and fibres produced in mechanical pulping of wood for paper-making). Dust particles can generally be quite easily separated by gravity or under the influence of electrostatic forces.

2. *Temperature and physical condition of people*: Temperature has an influence on the solubility and on the chemical activity of products. The relationship is not linear but exponential: its form is $e^{1/T}$ where T is the absolute temperature. As the temperature range for life on earth lies between about –20°C to +40°C, the influence of temperature on the product toxicity within these limits may be considered only slight. When the temperature rises, the effect is greater due to some secondary effects that either increase the product concentration in the system or affect the resistance of the organism. It may also happen that a person's resistance may have been reduced by

previous contact with specific products or by illness with the consequence that a toxic product could have a harmful effect at a lower concentration. For instance DEHP used as plasticizer in plastics affects such people far more than healthy ones.
3. *Potential interactions between components*: A particular medium (air, soil or water) may become unsuitable for life if some products, each of a low toxicity, can by secondary reactions generate more toxic products. Conversely, secondary reactions may also have favourable effects, e.g. when insoluble compounds are created, reducing the material in solution. However, the problem is usually only shelved since, finally, the new insoluble products will deposit on the river bottoms with other sedimentary particles and disturb the normal mud. Again, issuing a licence to discharge requires a thorough environmental survey.

b) Temperature

Temperature or heat is an important ecological parameter. Many chemical and physical phenomena are linked to temperature; the viability of habitats for plants and animals is also determined by the temperature. The aquatic biotope deserves special attention because the discharge of hot water into the rivers is no longer negligible (limited solubility of oxygen) since electricity has been produced on a large scale.

c) BOD – quotient

BOD (Biochemical Oxygen Demand) is the amount of oxygen required by the biological processes of breaking down (oxidizing) organic products under standard conditions (5 days at 20°C). The BOD-quotient of effluents, expressed in mg/litre, also measures the amount of biologically degradable organic compounds present in an aquatic biotope. Existing microorganisms (bacteria, moulds, etc.) will ingest such products as food and will grow rapidly. As the reproduction of microorganism populations is already fast (three generations per hour), the oxygen consumption for growth and processing in the existing culture medium is important. The oxygen present in water is used preferentially by this kind of waste treatment to the detriment of the higher aerobic organisms, causing a fall in their life expectancy.

At first sight, a paradoxical situation seems to have been created for biological life in water since an excess in food leads to a reduction of the expectation of life. In fact, variations among the microbiological species and their populations tend after a certain period to return to the original ecological equilibrium. If after discharge of waste water, repair mechanisms are disturbed or can only react at high dilutions, then we have pollution which should, of course, be avoided.

Similar mechanisms can occur in soil and air: although atmospheric oxygen can usually not be considered as a 'missing' element, the noticeable vanishing of higher animal species because of the detriment of lower ones (e.g. birds after ingestion of contaminated mussels) is a direct consequence of toxic products and of their accumulation in the food chain.

d) Acidity (pH)

The content of acid or basic components is a major ecological parameter for life in water and soil. Gaseous products present in the air such as hydrochloric acid (HCl) and sulphur dioxide (SO_2) which form acid solutions with water, will be absorbed from the air because of their solubility and/or reactivity to water and will eventually end up in the rivers and on the ground.

In the same way that temperature extremes limit the possibility for life, so limits also exist for acidity. Expressed as pH concentration they range from 4.5–9.5. Water is neutral when its pH is 6.5–7.0. Outside these limits, life in water becomes precarious for most organisms: the structures of the vital proteins and enzymes are always modified.

The pH control of waste water is therefore very important for the protection of the environment: during the different stages in the production of packaging materials, acids (sulphuric acid, nitric acid etc.) and bases (sodium hydroxide, calcium hydroxide, ammonia) are often used. In Belgium according to a Royal Decree, the pH of effluents from these chemical sectors must lie between 4.5 and 9.5.

e) Osmotic pressure

This parameter is also important for organisms in direct contact with an aquatic biosphere, such as microbes, fish, white and red blood cells, plants, etc.

Elements dissolved in water induce a so-called 'osmotic' pressure that can be calculated by Van't Hoff's law derived from Boyle-Gay Lussac's equation for gases:

$$\pi = \frac{f.g}{\{MM\}} \cdot \frac{1}{V.RT}$$

where: π = osmotic pressure (in bar)
f = correction factor for imperfect solution
g = weight of the dissolved element (in gram)
MM = molecular mass of the dissolved element
V = volume of water (in litre)
R = gas constant (0.082)
T = absolute temperature (t°C + 273)

The osmotic pressure is the pressure the dissolved element would exercise as a gaseous element in a volume V at the specific temperature.

Osmosis is the phenomenon by which the molecules of the solvent – in the case of biological processes it is water – diffuse through a semi-permeable membrane in one preferential direction determined by the concentration and the nature of the dissolved substance. Osmosis is very important because the cell walls of bacteria, plants and animals are thin membranes which water molecules can pass through more easily than dissolved molecules. When the aqueous solution exerts the same osmotic pressure on both sides of the cell membrane, the osmotic effect is not noticeable. Such 'Isotonic' solutions provide an exchange of solvent that is the same in both directions. Cells with a solution of lower osmotic pressure (hypotonic solution) will swell and burst. Conversely, they will shrivel in a solution of higher osmotic pressure (hypertonic solution). It is obvious that the moisture needs, and thus the viability of plants and animals, depends on the osmotic pressure of the aqueous medium with which they are in contact. Too often, restrictions on dissolved waste materials are expressed as a maximum concentration (gram/litre) whereas molar concentration should really be taken into account. The osmotic pressure of a 0.97% kitchen salt solution (7.7 bar, isotonic with blood plasma) is almost equivalent to that of a 10% sugar solution. This demonstrates that, besides the actual toxicity of dissolved products, the biosphere may become unsuitable for life because of the difference in osmotic pressure. As the molecular mass is part of the denominator in the equation this effect will be so much faster than when this term is low.

f) Influence of ionizing radiations

Since the 1950s, the danger of exposure to ionizing radiations has increased due to the development of nuclear power. There are many radiations with varying effects on materials. For instance the heating effect of solar infrared radiation comes from friction by molecular vibrations it generates in the material. The higher the energy content of the radiation, the greater the effect can be. Ionization causes electrons to leave atoms and, when the energy is high enough, this effect can induce modifications in the nature or structure of the irradiated molecules.

Photosynthesis uses radiations at the lowest energy level still able to provoke ionization. All other radiations have a higher energy content and may be classified as 'corpuscular' and 'electromagnetic'. Corpuscular applies to particles from the atomic nucleus [namely α-particles (helium nuclei), protons (hydrogen nuclei), β-particles (electrons) and neutrons] which can be emitted at very high speed naturally by 'radioactive' elements.

The radiations we can see (light rays) and the invisible x-rays (Rontgen rays) and γ-rays, are typical examples of parts of the electromagnetic spectrum. To express the quantity of energy absorbed either by a material or especially by organisms, we use the 'Rad' unit (abbreviation of Radiation Absorbed Dose) equivalent to 0.01 Joule per kg. According to the 'International System of Units', the Rad has now been replaced as unit by the Gray (Gy): 1 Gy = 100 Rad = 1 Joule /kg.

The 'Rem' (abbreviation of Rontgen Equivalent Man) is a safety unit indicating the biological impact of radiations on human beings.

The influence ionizing radiations could have on living organisms is of two sorts:

1. These radiations can attack and/or destroy cells and/or vital functions of the organism and thus directly limit its life expectancy. For sterilization purposes, foodstuffs (e.g. potatoes) and packaging materials (e.g. plastic pouches for blood) may be exposed to ionizing radiations and surface micro-organisms will be killed. This can also stop the germination of cereals, seeds and potatoes. It is obvious that complex organic molecules, components of foodstuffs, can also be modified by irradiation depending on the dosage. The doses allowed by law must not induce any noticeable modification in the food value of foodstuffs; however, some components, fats for example, are easily broken down with the consequence that irradiation could change the taste (166).

Sterilization processes using ionizing radiations and treating goods through their packaging, can produce structural changes in some packaging materials, especially plastics. Significantly increased migration has been found when examining irradiated polyethylene and polyvinylchloride packaging films (113, 75). The derivatives produced by the ionizing radiations are termed 'adventitious additives' (i.e. 'non-intended additives'). Like 'intended additives' they can pass into food by migration; their toxicity is not yet accurately known or determined (see also toxicity of packagings page 207 *et seq.*).

2) Ionizing radiations can cause abnormalities through mutations induced in the genetic code of individuals. The nature of these modifications is generally such that they damage specific characteristics, capacities and appearance. Since, the danger of ionizing radiations consists mainly in the dose absorbed over a certain period of continuous exposure,

the maximum allowable dose within a human lifetime is limited to 250 Rem: i.e. a maximum of 5 to 12 Rem in one year. In this respect, radioactive elements are a permanent threat; they can never be 'chemically' separated from the similar non-radioactive isotopes; e.g. radioactive phosphorus nuclei will be absorbed by the living matter and inserted in cell structures just like the non-radioactive ones so that the organism – or at least a part of it – can be exposed to permanent irradiation. Protection must be provided for the handling of all radioactive elements during extraction, enrichment, energy generation (in nuclear power plants), recovery from nuclear waste and in waste disposal in order to avoid direct exposure to radiations and the dispersion of radioactive elements into the environment.

g) Influence of noise

Finally, let us turn our attention to noise pollution, more especially since the choice of a packaging material can help limit the noise inconvenience in the packaging operation. Noise arises whenever vibrations in the air are propagated and perceived as fast successive pressure variations. The intensity and frequency of such vibrations are the parameters controlling this nuisance. The consequences of noise on humans are twofold: first, the threshold of audibility will be affected if people stay continuously in noisy areas; noise with a high energy content can damage the eardrum and cause deafness. Secondly, noise will result in more rapid fatigue and can create stress physiologically perceptible as increased blood pressure and digestive troubles. Noise will generally result in the lower efficiency of workers due to the fact that their attention is disturbed. Workers can be protected legally: according to the 'Occupational Safety and Health Act' (USA – 1970, 104, 106), a maximum exposure period in noisy surroundings should not be exceeded. Moreover, where there is daily exposure to various noise sources, the sum of the relative exposure periods compared to the maximum allowable periods should not exceed 1.

The calculation is as follows:

$$\frac{t_1}{T_1} + \frac{t_2}{T_2} + \ldots\ldots\ldots \frac{t_n}{T_n} \leq 1$$

where t_1 to t_n = actual exposure periods

T_1 to T_n = relevant maximum allowable exposure periods for the various noise levels

Industrial noise arises mainly from impacts and friction between objects, from turbulent circulation of liquids and from fast expansion of gases (e.g. internal combustion and jet engines). It is also well known that hollow containers such as glass bottles and metal boxes, rattle during displacements on conveyor belts. The packaging sections of dairies and breweries are striking examples. On the contrary, plastics used for the same packaging operations are softer and provide a significant noise abatement effect. This applies also to paperboard packaging.

III
Packaging materials

1 INTRODUCTION

This chapter aims to highlight the nature and variety of the main groups of packaging materials. The objective is to describe briefly how these materials are produced from their respective raw materials and to specify the possibilities in recycling. Where applicable, we will finally consider the hygiene factors in the industrial production of packaging materials. This is not intended to cover the whole field of all possible package forms and sizes for each material. However, we will pay particular attention to the basic principles; a survey of the applied processes will outline the energy and material consumption as well as any environmental problems.

Earthenware (pottery), wicker baskets, textiles and leather used for bags will not be studied: firstly, because the material consumed in these sectors is less important and secondly, because their production are crafts rather than industry and the problems relative to the environment and waste treatment are very minor. The procedure we have adopted to examine the materials is mainly based on their recycling possibilities. For comparison, the same sequence is also adopted in studying the energy aspects in the next chapter.

2 GLASS

a) Historical background

It is generally agreed that the use of glass was derived from the art of pottery: first, the glass was used to glaze terracotta beads and amulets. Green glazed beads from the period 12 000 BC and sky-blue amulets in glass dated 7000 BC have been discovered in Egypt (146). The fine sand needed for their production was found near Alexandria; the alkaline compounds required came from the 'salt lakes' located in the western part of the Nile delta.

In 1550 BC, glass-making was already very prosperous in Egypt: bottles and ornaments were produced in various colours. However there was no plain glass. As far as we know, the oldest piece in plain transparent glass is a vase found among the ruins of Nineveh; there are some inscriptions on it about the Assyrian King Sargon from the 8th century BC. This precious historic piece is in the British Museum.

The use of the 'blowing iron' to shape glass may have been invented by the Phoenicians about 300 BC. From the creation of plain glass during the 1st century to its industrial production in the 18th and 19th centuries, 'glass' remained an expensive product with specific applications (e.g. window glass for stained glass windows in cathedrals).

The supremacy held by the Venetian glass art over the rest of Europe passed to Bohemia during the 18th and 19th centuries. Modern techniques reflect the old glass art: the present manufacturing methods differ from the old ones only in their degree of automatization; blowing irons are still used but they are now connected to compressed air distribution systems, and the handling is automatic. The simultaneous use of several automatically controlled blowing irons and presses, reach outputs of approx. 100 000 units/day. Hence, glass has become a much less expensive everyday product.

b) Definition

The 'glassy' condition is not limited to what we usually call 'glass'; in fact, many natural resins (e.g. amber) and synthetic materials (plastics) feature a similar 'vitreous' state: the material appears in a non-crystalline frozen liquid state, with a viscosity so high at room temperature that it behaves as a 'solid' material.

Chemically, container and window glass is a homogeneous solution obtained by fusion of different metallic silicates into one or more vitrifying coagulated non-metallic oxides.

c) Components and raw materials

The main silicates used are associated with the following metals: potassium, calcium, barium, lead and aluminium. Generally these metals are not supplied to the glass furnace as silicates but as carbonates, sulphates or oxides which react at high temperature (ca 1460°C) in the 'glass furnace' with quartz sand (silicon dioxide) to form silicates. For clear glass, sodium and calcium carbonates are used. Unlike the calcium carbonate (occurring as marble and limestone), the sodium carbonate is not quarried but obtained from limestone and common salt (sodium chloride) by the addition of ammonia (Solvay's process): hence sodium carbonate is the most expensive component in glass production.

The main non-metallic oxides used are silicon dioxide (sand) and boron trioxide. Quartz is a pure silicon dioxide. Table 3.1 gives the relative content of the various oxide components in some glass products. Sand is always used and the other components influence the physical and chemical properties of the melt: calcium, sodium and potassium silicates lower the softening point; the presence of barium oxide increases the diffractive index and resistance to chemical attack. The addition of lead compounds to the melt improves the diffraction capacity and significantly improves the glitter (crystal glass); boron compounds, even at quite low concentrations, will considerably reduce the coefficient of expansion of glass. Among the range of glass-like products, quartz has the lowest coefficient of expansion (5.10^{-7} per degree instead of 33.10^{-7} per degree and 90.10^{-7} per degree for Jenaglass and soda glass respectively).

Generally, iron and chromium oxides are undesirable, as they colour the glass yellow to brown or green respectively. For the preparation of colourless glass, these components, whether they are present as impurities in the raw materials or come from the cullet (recycled glass), should be strictly avoided. Other glass colourants are manganese dioxide to give a purple, copper oxide for turquoise and cobalt trioxide (Co_2O_3) for blue. Small gold chips will give glass a ruby red colour while black coloured glass is obtained by the addition of both manganese and cobalt oxides.

d) Manufacture of glass packaging

After the raw materials have been powdered, they are accurately weighed and thoroughly mixed. At the entrance to the glass 'furnace' (Figure 3.1), broken glass (cullet) and the raw materials mixture are fed in. The latter melt in the cullet and at the high temperature inside the glass furnace the carbonates turn into oxides which then react with sand to form silicates. For instance, the following reactions apply in the case of calcium carbonate:

$$CaCO_3 \xrightarrow{t} CaO + CO_2$$

$$CaO + SiO_2 \longrightarrow CaO.SiO_2$$

The breakdown of carbonates generates carbon dioxide gas in the hot glass melt which assists in good mixing of the components. Moreover, a mass deficit occurs to the detriment of the supplied raw materials: there is a 44% drop from the calcium carbonate and 42% from the sodium carbonate due to the loss of carbon dioxide. Combustion gases (water, carbon dioxide, etc.) escape through a chimney to the atmosphere. In the second (non-heated) part of the furnace, the glass mass cools but, while still soft, is delivered to the bottle or jar blowing process. Figure 3.1 indicates how a bottle is shaped in a mould by means of a 'blowing iron'. Unlike rolled glass (e.g. tubes, bulbs) or glass pressed in an open mould (bowls), bottles and jars have a small longitudinal and continuous bead on the external surface corresponding to the narrow gap where the two halves of the moulds come together. With the glass mass, various packaging wares can be shaped in closed moulds

Table 3.1 Main components (in %) of glass products

Components	Non-metallic oxides		Metallic oxides					Others
	SiO_2	B_2O_3	CaO	Na_2O	K_2O	Al_2O_3	PbO	
Glass products								
Quartz glass:	100	—	—	—	—	—	—	—
Hollow ware for packaging:								
Soda glass (Bohemian glass)	76	—	7	3	14	—	—	—
Yellow-green glass (bottle glass)	65	—	10	12	3	6	—	$2Fe_2O_3$ 2BaO
Blue-green glass (bottle glass)	65	—	10	12	3	6	—	2FeO 2BaO
Green glass (bottle glass)	68	—	10	12	6	—	—	1 Cr_2O_3 3 BaO
Brown glass (beer bottle)	64	—	10	12	3	6	—	$1Fe_2O_3$ 3 Carbon
Lead glass: crystal glass	65	—	2	—	15	—	18	—
Flat glass: window glass	73	—	10	13	1	1	—	2MgO
Laboratory-ware Thüringer glass	70	—	8	12	6	4	—	—
Jena glass	74	5	1	8	—	8	—	4 BaO
Pyrex	81	11	1	5	—	2	—	—

[Source: Fink (52); *Encyclopedie der Materialenkemis* (184)]

by the blowing iron method: we can find large or small, cylindrical, flat, square or hexagonal wares, well decorated (or not) on the market today. However, all of them feature a circular opening, the standard diameter of which depends on the nature of the product to be packaged. The finish of the bottleneck is designed to fit the chosen sealing system, either a threaded closure, a crown cork, or the earlier bark cork. Various alternatives to these, such as 'snap-on caps' for pharmaceuticals, aluminium foil for bottles of milk, etc., can all be provided with an

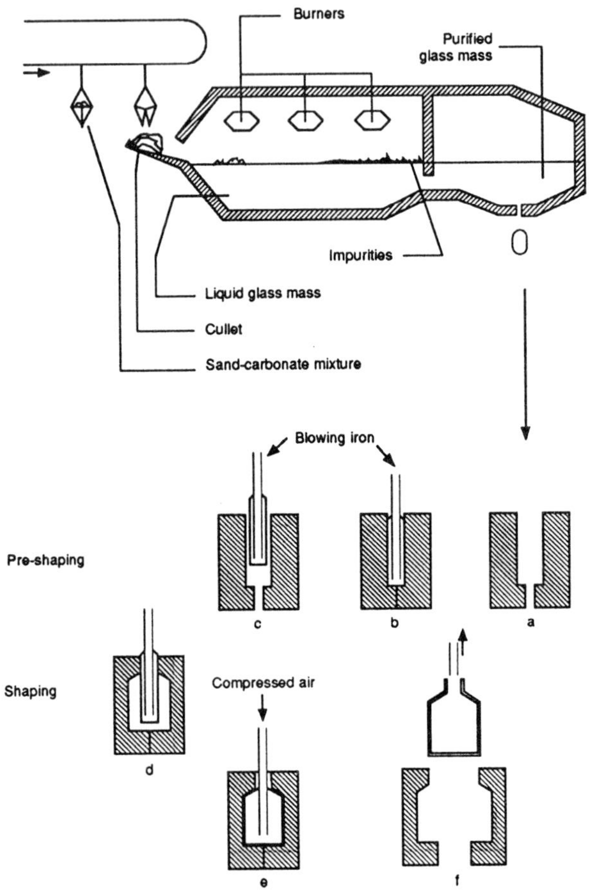

Figure 3.1 Sketch of a glass furnace and shaping of a glass bottle

appropriate sealing system which breaks on opening. It is also obvious that the environmental impact is limited. Broken glass (cullet) is needed as 'raw material' for glass-making. Recycling of glass, therefore, whether separated from household refuse, or collected separately from home or in bottle banks, needs a sequence of cleaning operations. Glass from household refuse is washed, sorted by colour, crushed and finally any metal pieces are removed.

e) Main characteristics of use

Glass packaging is mainly used for drinks (including alcoholic and carbonated), foodstuffs, pharmaceuticals and chemicals. In clear glass, the contents remain visible (an emotional element) but on the other hand any light sensitive components (vitamins etc.) are more rapidly decomposed and lost. Ultra-violet and visible rays can pass through quartz glass, but the more common clear glass bottle acts as a filter for UV-rays, and coloured glass (brown or green) will stop most visible rays (see page 38). Glass is chemically inert, inodorous and impermeable to liquids and gases. With the exception of ampoules, glass packages can be re-closed after opening which is a major requirement for their use as returnables. Glass packages can also be sterilized by all available methods. They have the disadvantages of fragility, heavier weight and rather limited resistance to local temperature variations. Due to the poor heat conductivity of glass, heat induces thermal stresses likely to cause breakage (thermal shock). Waste glass (cullet) is rigid, dense and incompressible, but completely recyclable.

3 METAL PACKAGING

a) Historical introduction

Iron, tin and aluminium are the metals mainly used in the packaging sector. Some other metals are used in alloys (e.g. tin with zinc and copper) or as solder (tin and lead). Metal packaging is usually made out of sheets from which rectangular or cylindrical containers can be shaped.

Ancient civilizations used the noble metals (gold and silver) which, because of their high value, were reserved for packaging expensive goods such as jewellery. The Romans knew how to roll soft leadsheets down to very thin foils. There is no clear evidence whether this 'toxic' material was for packaging, but it is clear that lead has been allied to other metals such as tin. Tin foil (actually an alloy containing some toxic components such as antimony, lead and copper) was used until about 1930 to package cheese (146). It was known popularly as 'silver'-paper.

Because of the rusting phenomenon, iron has not been used in direct contact with foodstuffs. The protective properties of a tin layer on iron were discovered around 1200 AD in Bohemia: the iron plate obtained by hammering was immersed in a bath of molten tin. This method was kept secret for approximately 400 years. Cans were first manufactured on an industrial scale by hand at the beginning of the 19th century (1810) at a rate of 20 pieces per day, per worker. Nowadays, steel plate has replaced iron; the development of better alloys and technologies makes it possible to get thinner rolled plates and thinner tin protective coatings. Automatic units can produce 400 to 1600 pieces per minute. The development of new synthetic resins for metal protection has led to the use of plain steel plate for many packaging purposes: so tinned steel and tin-free steel, with or without a protective layer of varnish, serve to package alcoholic, neutral and acid beverages, foodstuffs, lubricants, chemicals, etc.

Aluminium, a packaging material widely used nowadays, was only discovered by Oersted as recently as 1825. Extraction from the aluminium ore bauxite was quite an expensive process, hence aluminium was at first considered as almost a 'noble' metal and its use was very limited: for instance, at the court of Napoleon III, cutlery made in aluminium was reserved for special guests. Heroult and Hall discovered the electrolytic process of producing the metal in 1886, and in 1888 Bayer devised the 'cheapest' method (treating aluminium oxides) and liberating the metal by electrolysis. The changes in the price of aluminium have been quite astonishing: from $545 per pound in 1852, it fell to $11.33 per pound in 1885, to 57 cents in 1892 and to 14 cents in 1942. Today 1 kg of aluminium costs approximately $1.26.

Nowadays the manufacture of most metal packages is practically fully automatic, hence they are cheap and in common use.

b) Metallurgy and manufacture of metal packages

Metals used in the packaging sector come mostly from oxides or sulphides. To obtain the metal, sulphides are usually changed first into oxides by oxidation (reaction with the atmospheric oxygen). This process generates sulphur dioxide which can be used for the synthesis of sulphuric acid. The metals can then be directly separated from their oxides by the so-called high-furnace process: at the high temperature obtained in a furnace by the combustion of coke, these metallic oxides react with the carbon, supplied by the coke, to form the metal and carbon dioxide. The liquid metal collects at the bottom of the furnace. The high-furnace gas also contains some combustible components (namely hydrogen, methane and the very toxic carbon monoxide) as indicated in Table 3.2. Because of its toxicity, but also its calorific content (900–1000 kcal or 38–42 MJ per m^3), high-furnace gas is not released into the atmosphere but burnt for plant heating (heat recovery). Metal production processes can pollute considerably, especially if dust clouds from the refining process escape into the atmosphere.

Aluminium, unlike iron, lead and tin, cannot be obtained from its oxide by the high-furnace process. After purifying, the bauxite ore is treated with concentrated soda which dissolves aluminium oxide but not the other impurities (iron oxide) that can be removed by filtration. The main impurity is iron oxide, which forms a red mud; this is very difficult to do anything with and currently produces what is known as 'red mud lakes', tending to pollute the land considerably. There is much research going on to deal with this red mud. The solution of bauxite in concentrated caustic soda is diluted with water, providing a deposit of purified aluminium oxide. After drying, this

Table 3.2 Typical composition of high-furnace gas generated in iron processing (approx. $6m^3$ per kg iron ore)

Non-combustible components %	Combustible components %
Carbon dioxide gas 10–16	Hydrogen gas 0.5–4
Nitrogen gas 52–60	Methane 0–3
	Carbon monoxide 23–30

is mixed with cryolite (sodium-aluminium fluoride) which lowers the melting point of the oxide significantly from about 2300°C to approximately 1000°C. From the molten 'salt bath' thus produced, almost pure aluminium is obtained by electrolysis.

Recovery of metals from domestic waste is facilitated by the magnetic properties of iron which enable it to be easily separated from other refuse. The other metals, not containing iron, can be separated by means of vibrating screens. The metals tin and aluminium can be recovered from discarded packages as follows: tin can be removed from tinplates by chemical processes and then precipitated. To recover aluminium, it can be remelted with no great problem, during which process most of the impurities come to the surface of the molten metal and can be skimmed off. Recovery of iron from scrap-iron by melting alone is not feasible, but the scrap-iron can be added to a blast-furnace with iron ore (Martin Siemens' process), the tin having been separated from the cans beforehand.

Table 3.3 lists the various raw materials for the production of each metal and reveals that in Belgium imports of iron ore, tin and aluminium are necessary. Taking into account the cost of energy and of raw materials used, metal recycling is economically worthwhile.

The manufacture of metal packages is in principle an easy process: starting from a rolled thin metal sheet, eventually with protective layers such as tin and/or varnish, pieces are cut, formed and combined together. As mentioned before, the tin layer can be obtained by immersion in molten tin; this cover can also be applied to the steel plate by electrolysis from an aqueous solution of tin salts (electroplating). With these processes, the protective layer always has the same thickness on each side of the steel plate, hence the quality of a final package can be judged by examining the external side only. On the contrary, only one side of the plate will be varnished. 'Tinplate' is still the most common raw material for tins and cans. Black plate is not often used and can be described as tinplate without the tin, i.e. mild steel sheet, available in the same range of thicknesses as tinplate. It is usually coated with a varnish. 'Tin

Table 3.3 Ores processed to obtain the principal metals used in the packaging sector

Metal	Ores	Iron content %	Origin
Iron	Magnetite (lode stone – Fe_3O_4)	60–70	Sweden, Norway, USA, Ural Mountains
	Haematite (red iron ore – Fe_2O_3)	50–70	Great Britain, North Germany, USA, Spain, Newfoundland, India, Morocco, Ukraine, South America
	Siderite (spathic iron ore – $FeCO_3$)	30–35	West Germany, Great Britain, France
	Limonite (brown iron ore – $(Fe_20_3)_2(H_2O)_3$)	30-50	Sweden, Finland, Spain, Lorrain, Luxemburg
	Pyrites (yellow iron ore – FeS)	60	Spain, Portugal
Tin	Cassiterite (tin stone – SnO_2)	–	Malayia, Bolivia, Thailand, China, Indonesia
Aluminium	Bauxite (Al_2O_3)	up to 30	Surinam, British Guyana, USA, France, Russia, Indonesia, Jamaica

free steel' is like tinplate but the tin coating is replaced by corrosion resistant metal such as chromium. Square and cylindrical shaped boxes can be made from these iron-based metallic sheets. Aluminium sheet is also used in increasing quantities.

Aluminium foil is also used to wrap foods and to produce trays. However, the image of metallic packaging has been projected by canned food. Until about 1970 the three-piece soldered tinplate can was more or less standard. Since that time considerable development has taken place. The construction of the three-piece can, however, has not changed fundamentally in over 150 years: a flat rectangular sheet of tinplate is formed into a cylinder and the join soldered to form a side seam. Circular discs of tinplate are then mechanically secured to flanges made at the end of the cylinder by a rolled seam, known as 'double seam'. One of these ends is fitted by the can-maker and the other by the packer after filling (see Figure 3.2).

Advances have been made in techniques but not in principles. Since about 1985, solder has been eliminated almost everywhere and electric resistance welding is used; laser welding is a prospect for the future. Automation has increased the speed of manufacture and better, more secure mechanical joins have been developed. The improvement of processing techniques has also resulted in a spectacular weight reduction of both the materials used for making cans for food packaging, iron and tin. In the period 1824 to 1990 the wall thickness of the steel plate decreased by about 67%, while the tin coating could be reduced from 13 to 0.2 microns in thickness (98.5% reduction).

Additionally, two-piece cans have been developed and today there are three major types of open top food and beverage cans in use. These are:

1. the three-piece welded can (see Figure 3.2),
2. the two-piece 'Drawn and Redrawn' – can (DRD-can), also made from tinplate,
3. the two-piece 'Drawn and Wall Ironed' – can (DWI-can).

The 'easy-open' ends; first for beverages and then with a full aperture for foods, are also being developed and these are now mainly made from steel as well as from aluminium (with a higher magnesium content), in which the developments first took place. Steel 'easy-open' ends are becoming more popular now that the initial problems have been overcome.

Surveying these more recent technical developments shows that the forces which bring them into being are:

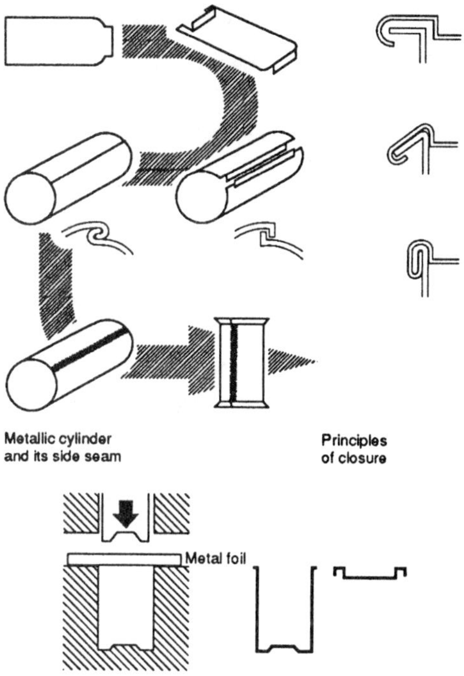

Figure 3.2 Manufacturing sequences for a three-piece and a two-piece can

1. the desire to reduce costs by using thinner plate (less metal is used, also resulting in waste reduction),
2. lower processing costs by the use of automation and higher speed production,
3. the provision of convenience for the user (easy opening systems, etc.),
4. better presentation: e.g., new shapes, better printing and graphics, and the use of metal plastic combinations (coatings, varnishes).

c) Uses of metal packaging

Metal cans have the advantage over glass in being of lighter weight. They offer good protection against the influence of light and, when closed, will resist the internal pressures of aerosols and carbonated drinks as well as the pressure generated by heat when foodcans are retorted. Aluminium foil is also a useful material and has become available in various thicknesses ranging from 20 μm (for combination with paperboard, paper and plastics) to 80 μm (for dishes, trays etc.). Very thin foils often have microscopic holes (pin holes), but since most foils are coated with a varnish these holes are sealed with plastic and do not permit moisture vapour or oxygen to enter sealed packages. Aluminium foil has a different appearance on each surface: one surface is matt, the other one bright. This is because the last rolling to make the foil very thin is made by calendering two sheets together where the surfaces on the inside become matt and the surfaces in contact with the rollers are brightened. Practically, apart from appearance, there is little difference in the two sides.

d) Resistance of metal packaging to corrosion

Oxygen and moisture are the principal factors needed for the corrosion of iron, and both must be present at the same time. The corrosion product of iron (rust) is an oxide; moisture is necessary for its formation and the metallic ions which are produced then dissolve with subsequent corrosion. Neither dry oxygen nor oxygen-free water will cause any appreciable corrosion of iron.

Tin has the property of reacting superficially with atmospheric oxygen to form a layer of tin oxide which prevents further corrosion in an aqueous or weak acid media. The brown iron oxide (rust) does not protect iron against further corrosion, hence the need for tin coating on cans. For most acid fruit juices, neither the tin oxide layer on tinplate nor the aluminium oxide on aluminium packages provide an adequate protection. Both oxide layers can react with acids from the package contents, liberating metallic ions in solution. These can change the taste and colour of many foodstuffs (122). The swelling of cans after the generation of hydrogen gas and the formation of holes result from an electrochemical reaction between the metal surface and the aqueous contents, as will be discussed further on. This happens with tinplate when the tin layer is damaged and the underlying iron exposed.

Billions of tinplate containers are successfully used each year for the packaging of a wide range of products; nevertheless many different sorts of attacks on the containers by the packaged products can occur. Many products will attack tin itself, but removal of the tin is considered preferable since dissolution of the tin by the product will delay an attack on the underlying steel. This is very important, since the tin coating is a very thin one and is in any case not 100% complete, particularly after the tinplate has been assembled into a can on high-speed equipment. If the exposed areas of the steel were not protected by the tin, this would lead to pin holes developing rapidly in the container. Attack on the tin, however, is also detrimental, because in many instances it causes the product to discolour or produce off-flavours due to metal pick-up. Blackening of the tin itself is the result of a chemical reaction of the tin with sulphur-ions from the packaged product (e.g. in tomato juices) to form black tin sulphide, which is insoluble. Hence in most cases the tinplate is also lacquered.

Where lead solder is present, the product may also attack this: highly alkaline products will even dissolve lead solder with consequent leakage at the side seam. Some organic compounds also react with it, resulting in the blackening of the solder and also the product. These problems can be reduced by lacquering the inside seam, but perfect coverage is difficult to obtain, so high tin solders are often used where problems would occur with lead solders.

e) The mechanism of corrosion

Corrosion is an electrochemical phenomenon which involves the transport or the displacement of electrons. If two different metals are immersed in slightly acidified water, and then connected, a current will flow due to the galvanic cell-reaction. Indeed, a specific electric potential is set up between each metal and the solution dependent upon their solubility. The two metals in such a cell are called respectively anode and cathode.

One of the important factors controlling the electrochemical behaviour of tin is oxygen. Normally, the concentration of oxygen in the product in a tinplate container is very low: under these conditions the tin is anodic and the steel cathodic. This is the desirable condition, because here the tin would be attacked by preference and thus protect the steel base, even if the tin coating contains pin holes. At higher concentration, however, the tin becomes the cathode and the steel the anode, and under these conditions rapid corrosion of the steel takes place, usually causing the tinplate to be perforated. The relationship between the corroded can and the galvanic cell consisting of iron and tin is schematically shown in Figure 3.3.

At the bare metal surface, the following reactions occur, provoking the swelling of cans by the hydrogen formed and the formation of holes by the dissolution of iron. The positive pole of tin receives electrons neutralizing the hydrogen ions to hydrogen gas:

$$2H^+ + 2e \longrightarrow H_2$$

the negative pole of iron liberates electrons through the transformation of iron into iron ions:

$$Fe - 2e \longrightarrow Fe^{2+}$$

As long as the two metals remain in contact with the liquid, the reaction will continue. Therefore buckled or damaged cans must be withdrawn from the market in Belgium according to the RD dated 15 March 1982.

The dissolution of metals such as tin by chemical or galvanic reaction is much reduced by varnishing one side of the metal plate with an additional protective lacquer coating (epoxy resin) before shaping into cans. Two-piece cans, however, are varnished on the inside after the drawing process by which they were formed. The probability that the holes in the lacquer will coincide with the pores of the tin layer is low.

Of course, the above-mentioned reactions are not limited to

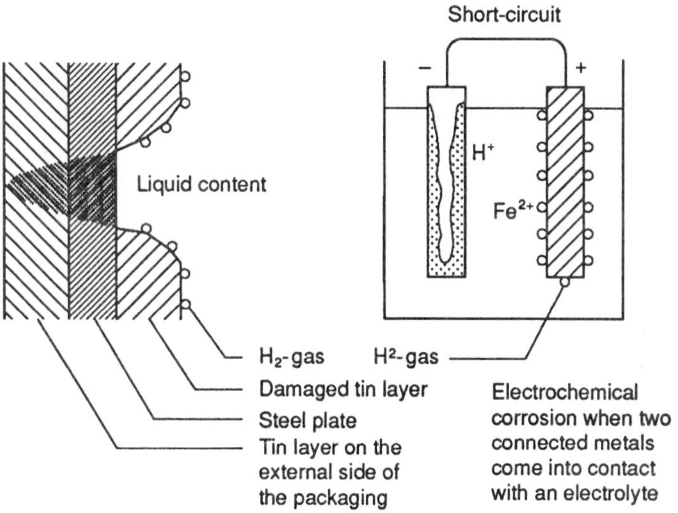

Figure 3.3 Parallel between a damaged tinplate and a glavanic cell made from tin and iron

metal packaging: in the kitchen we also use pans, dishes and cutlery made of different metals and/or alloys. Electrochemical reactions will always take place when different metals are joined to each other in an aqueous medium; for example, aluminium foil will be corroded when used to wrap a silver dish containing moist foodstuffs such as boiled potatoes. Various metals coming into contact with each other during washing-up (e.g. in fully automatic dish-washers) will also be corroded and typical, slightly coloured stains will generally appear on their surfaces.

The generation, during cooking or preservation of foodstuffs, of undesirable metallic ions in excess from cans can thus be limited by avoiding contact with other metals (cf. Chapter VI and Table 2.3).

Thus metal packages are eminently suitable for the preservation and transportation of many goods, as they can be manufactured and protected against corrosion by modern techniques.

4 PAPER AND BOARD

a) Historical introduction

The progress of paper and its technology has probably been mainly supported by the human will to record its knowledge for future generations. During the 18th and 19th centuries, when the 'paper machine' was in common use, paper became a packaging material, although the use of thin sheets to wrap goods was already known in early times (such as the use of leaves, tree bark, thin metal foils). Even in the 5th century BC, the Egyptians manufactured paper, but it could hardly be used as packaging material because of its high cost. Originally, paper and 'papyrus' were designed as writing materials to replace parchment (treated animal skin) and vellum (treated skin of new born kids or lambs). Papyrus is made by weaving the long fibres obtained by crushing the internal soft stalk from the lower part of the papyrus plant. The gaps in the final woven material are filled with a fine powder (clay). The sheet is then coated with a natural wax as binding agent and after drying is rolled up.

Nowadays 'paper' is a thin flexible sheet consisting of a random network of small fibres, fillers and binding agents. Whereas long cellulose fibres form the basis structure of papyrus sheets, today the shorter cellulose fibres from wood and cotton are used in paper-making. The principle of the present paper-making process was discovered in China around the year 105 BC. Its inventor, Tsai Lun, was considered as the god of paper-makers.

Wet pulp produced from torn hemp, rope and the finely ground bark of the mulberry tree was used. During the 8th century, the art of paper-making was transmitted to the Arabs by Chinese prisoners of war and came, via Spain and Italy, to North Europe during the 12th century. In 1390 the first paper mill was established in Nurnberg (Germany). Machine-made paper has been produced since 1405 in the South Netherlands (in Hoei) and since 1586 in the Netherlands. The paper pulp was made in large wooden buckets with a fine wire mesh at the bottom (dip paper). The drained and dried screen content formed the paper sheet.

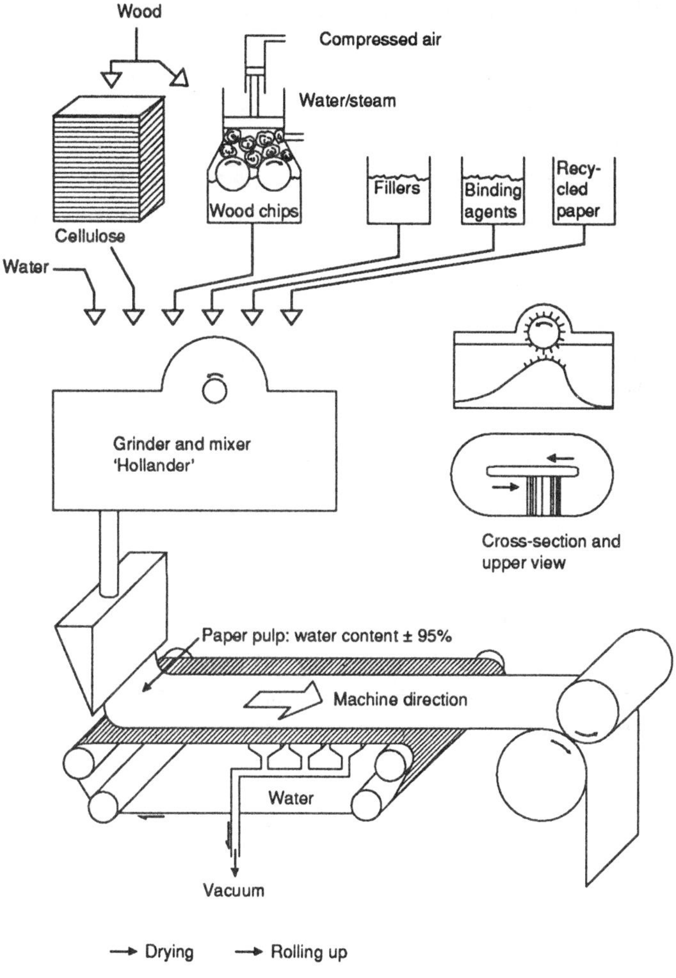

Figure 3.4 Diagram of papermaking

In the 16th century a steady desire for faster and wider cultural exchanges as well as the invention of the art of printing during the 15th century (Gutenberg, 1440) contributed to the development of the automatic paper machines which spread the paper pulp on high-speed screens (see Figure 3.4). This was the invention of the Frenchman Nicholas Robert in 1787 and was improved later by Bryan Donkin in England. In 1807 the invention was taken up by two London stationers named

Fourdrinier. They were relatively unsuccessful commercially and finally went bankrupt, but the type of machine used has been given their name. The modern Fourdrinier machine has eight basic parts – four comprising what is called the 'wet end' and four the 'dry end'.

In the further development of the paper industry during the 19th and 20th centuries, and in the design of modern high-speed machines which can produce 1150 tons/day, we still find the basic principle elaborated by the Chinese and the Ancient Egyptians. A dilute suspension of paper fibres is formed onto a rapidly moving wire mesh screen which carries the stock from the place where it is poured onto the wire mesh screen up to where the sheet is removed at what is called the 'couch' roll. The function of the wire mesh screen is to permit drainage of the water to take place, and this is held by rollers or suction boxes underneath the riddle. Once the mass of the mat of paper fibre is sufficiently strong, it can be transferred to a felt which moves it into the drying section. At this stage the mat of fibres has approximately 75–80% of water in it and it passes through presses and then through drying calenders which dry down to a reasonable level. The moisture content in general at the end of the paper machine lies between 6% and 8%.

Paperboard is produced by a similar process but is thicker and more rigid than paper. The dividing line is generally considered to be a basis weight of about 250 g per square metre of the sheet.

b) Environmental impact of paper-making

Raw materials and resources

Wood is the most important raw material for making the cellulose fibres (pulp) which are used for paper-making. Unlike some other raw materials for packaging, wood is a naturally renewable resource (Figure 2.1).

Approximatively one third of the Earth's surface is covered by forest (i.e. $4.1. 10^9$ ha out of $13.39. 10^9$ ha); 30% to 40% of these forests produce timber for one use or another. 50% of all timber is used as fuel. Wise forest management and reforestation can maintain the necessary forest reserves. About 28% of the timber produced is used in the packaging sector for wooden cases, crates, paper and board (72). In 1979, the world production of paper and board reached $173.8.10^9$ tons, of which $73.19.\ 10^9$ tons was used for packaging (42%). In Belgium in 1982, 1415 thousand tons of paper and board were processed; some of this was imported, some was produced at home. The total used for packaging was appproximately 611 000 tons (43%) divided as follows:

81 776	tons for paperboard boxes and cases (13.4%)
276 132	tons for corrugated board boxes and cases (45.2%)
77 754	tons for paperbags and sacks (12.7%)
33 620	tons for other paperboard packaging (5.5%)
78 258	tons for packaging in paper (12.8%)
10 974	tons for paper laminated with aluminium foil (1.3%)
52 500	tons was coated with plastics (8.6%)

(Source: AGEFI, February 1983)

Other raw materials which contain cellulose are the bark of the mulberry tree, cotton, straw, espartograss, flax, bamboo, bagasse (waste from the sugar cane) and waste paper; all are mainly used for making lower grade packaging papers and boards, although some are used to make printing and writing papers.

Pulping processes

Pulp preparation from suitable hardwoods or conifer softwood, may be done by three methods: mechanical, chemical or hybrid processing. All pulps may be used unbleached or bleached and are produced by various methods. 'Mechanical pulp' is produced by taking logs of soft wood, mainly pines, debarking them and pressing them against rotating grindstones with plenty of water. This mechanical pulp is a suspension of wood fibres in water (paper pulp); it is the main raw material for newsprint. Besides the cellulose matter (50% of the tree) there is also lignin (30% of the tree) present at the same time. This natural binding agent is the cause of the yellowing and wet strength of newsprint under the influence of light and oxygen.

'Chemical pulp' is made from wood which is first chipped and then subjected to the action of chemicals at a high temperature in a digester. The non-cellulose components (mainly the lignin) are chemically dissolved. In the sulphite process, the wood chips are treated with calcium hydrogen sulphite and the whiteness of the cellulose fibre is preserved. In the sulphate process, an aqueous solution of sodium hydroxide (also calcium hydroxide and other bases) is used. This method gives a brown coloured pulp which requires bleaching if it is to be whitened. The latter process is also applied to treat cheaper sources.

In the hybrid or semi-chemical process the wood chips which are usually from beech or birch trees are treated partly by chemicals and partly by mechanical grinding to reduce them to fibres, hence the name 'semi-chemical'. Semi-chemical pulps are generally used for the manufacture of fluting medium for corrugated board.

The next stage in the process is to beat the pulp to make it capable of binding together more effectively. The objective of the beating process is to rub and brush the individual cellulose fibres, causing them to split down their length in such a way as to produce a mass of thin fibrils which will enable them to hold together in the matted paper far more strongly. This process is called 'fibrillation'. The greater the degree of fibrillation that can be produced, the greater the strength of the paper. Different pulps respond differently to the treatment: softwood fibres fibrillate much better than hardwood or straw fibres. So softwoods produce the stronger or better papers. Beating also breaks up any clumps of fibres and reduces them to individuals. The art of beating is to maintain a high proportion of fibrillation and not to cut the pulp too much; this gives the desired properties in the finished paper.

At this stage – depending again on the desired paper quality – recycled paper (non-inky and treated waste paper), fillers (the main ones for which are clay, titanium dioxide, barium sulphate [permanent white], chalk and talc) and binding agents (glue, waxes) may also be added. The resulting homogeneous diluted suspension of fibres in water then flows through a slot, as a curtain, onto a wide, high-speed (up to 60 km/h) screen conveyor with a very small mesh size. A low underpressure beneath the screen sucks the water out of the pulp while the fibrous material with both the fillers and binding agents stays on the screen. This forms a wet but compact sheet which is then pressed, dried and glazed by a series of calenders at the dry-end of the machine. At the end of the process, the paper is mechanically wound. Modern equipments supply rolls in widths up to 8 metres, these rolls being cut afterwards to the desired dimensions, or to sheets to manufacture into packaging.

Nowadays paper bags are produced by fully automatic machinery: remote control by computerized devices commands the successive operations: cutting, folding, shaping, filling, sealing ... and bundling of unit packages, ensuring a good uniformity within the weight class and quality.

A typical composition of newsprint and paper for periodicals might be:

Material	*Newsprint (%)*	*Periodicals (%)*
Wood pulp (+ recovered paper)	70	52
Chemical pulp	17	17
Filler, coating and binding agent	5	25
Water	8	6

Papers which are designed to package foodstuffs do not use mechanical pulps but use the more purified material such as sulphite and sulphate pulps.

To produce cartonboard, several layers of pulp are deposited on a rotating screen conveyor to build up to the desired thickness. The methods of treating the fibres before the boardmaking process are virtually the same as those used in papermaking. After the sheet is formed, however, the methods of removing the water, pressing and drying, while similar in principle, differ in detail depending on the method used.

Corrugated paper and board differs from plain paper and board in the following way: corrugated board consists essentially of two flat parallel sheets, called liners, with a central fluted or corrugated sheet between them. The three sheets are held together with adhesive which is applied to the tips of the fluted sheet. The materials used for any of the three basic components can be in a variety of weights and types of paper and board, and the flute configuration can be varied. The use of this type of structure considerably increases the rigidity and the insulation capacity of the finished product.

Recovery and recycling of paper

As the main constituent of paper and board is fibrous, waste paper and/or used paper can be recycled. Fresh clean paper waste, not printed, obtained during production on the paper machine, can be directly repulped in the pulping process and thus recycled. However, printed papers (newspapers, periodicals) must first be treated to remove ink, clips, plastic rings, etc., which are present from their particular use. Unlike glass, paper and board cannot be recycled indefinitely: the cellulose fibres deteriorate by use, by the production processes and by potential de-inking treatments; the shorter fibres become so small that they will pass through the web and the longer ones are reduced in length. In order to maintain the quality of the paper or board, especially its colour and resistance properties, the amount of 'waste paper' recycled by mixing with the fresh paper pulp, is limited.

The amount of recovered paper used in paper and board production in Belgium reached 18% in 1977. This small quantity (the average in other EC countries was around 30% and in the Netherlands reached as high as 53.8%) was due to the fact that, in Belgium, approximatively 68% of the waste paper collected (243 000 tons out of 377 000 tons) was exported. Five years later the percentage recycled as against the total produced in the EEC and in Belgium changed, as shown in the following table (1982):

	EEC (%)	Belgium (%)
Printing paper	6	5
Corrugated fibreboard	74	97
Packaging papers	43	15
Carton boards	76	60
Others	31	29

[Source: Thomas, N.V. Papierfabrieken van België, 1982]

Paper produced from recovered paper only (100% recycling) is grey: further bleaching would destroy the fibrous structure and hence reduce the quality of the paper.

This short description of paper technology shows that environmental damage can occur especially during the pulping processes: in addition to the chemicals which are capable of polluting the environment (alkalis, sulphur compounds), the effluents from the digesters contain all the non-cellulosic components of wood and their derivatives. The acid (sulphite pulps) or basic (sulphate pulps) character of the reactants, together with the high concentration of organic matter, means that direct discharge of these effluents into rivers, etc. would severely pollute. Effluent from paper mills can also contain other pollutants: mainly some small cellulose fibres, fillers and resins which pass through the machine wire and flow away diluted by a large quantity of water. The impact of these effluents from the paper machine depends very much upon the raw materials used, whether it is mechanical or chemical pulp, etc. Hence the composition varies significantly as shown in Table 3.4.

Table 3.4 Characteristics of effluents from the paper industry before treatment

	Concentration (mg/l) before treatment	Maximal allowable concentration (mg/l) for discharge
Floating components	100 - 900	70
BOD	50 - 400	40
COD	150 -1300	360

Table 3.5 Main process data relative to the production of 1 ton unbleached kraft paper

Raw material		*Discharge into atmosphere (atmospheric pollution)*	
Wood chips	1120 kg	Sulphur dioxide	18.13kg
Auxiliary materials		Hydrogen sulphide	13.1 kg
Sodium sulphate	73.2 kg	Mercaptans	3 kg
Calcium oxide	11.3 kg	Solid particles	80 kg
Binding agents	10 kg		
		Water pollution	
Water	±100 ton	Calcium oxide	7 kg
		Sodium carbonate	1.59kg
Steam	10.3 ton	Sodium hydroxide	3 kg
		Sodium sulphide	1.15kg
Energy		Solid particles	38.4 kg
Fuel	54 kg		
Electricity	354 kWh		

Effluents must therefore be treated to meet the requirements for discharge (cf. RD 3 August 1976).

Table 3.5 is a digest of the major process data relative to the preparation of 1 ton of kraft paper (sulphate process) (15). Note the large quantity of water used: approximatively 100 times the weight of the woodchips. As already mentioned, the liquid pollutants vary according to the chemical treatment of the wood. De-inking installations for recycling waste papers generally use detergents: these will preferably react with ink during the flotation process and form a foam that can be removed. The resulting ink concentrate is toxic and must also be treated. The atmospheric pollution mainly results from the energy production (combustion of fuels).

Environmental pollution by industrial paper production has decreased significantly over the last 30 to 50 years due to greater care and various standards enforced by the authorities. In Sweden, for instance, discharges polluting the environment

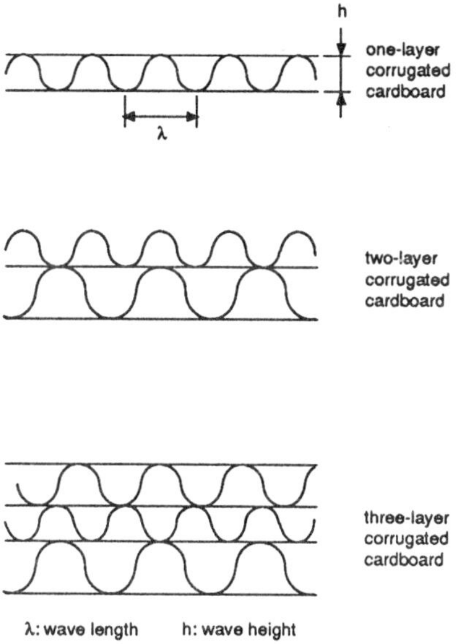

Flute configuration	Nr of flutes per metre	λ mm	Flute height h (mm)	Minimum flat crush (N/m^2)
A (coarse)	104-125	9.6-8	4.5-4.7	140
B (fine)	150-184	6.7-5.4	2.1-2.9	180
C (medium)	120-145	8.3-6.9	3.5-3.7	165
E (very fine)	275-310	3.6-3.2	1.15-1.65	485

Figure 3.5 Various corrugated fibreboards and their characteristics. [Source: Paine]

were reduced by almost four times between 1970 and 1980 and have been reduced even more since then. These data show that overall the paper industry is not as environmentally safe as it might seem at first, before effluent discharges were regulated.

c) Characteristics of use of paper and board

Paper and board qualities are characterized by their weight per unit of surface (grammage or basis weight). Tables 3.6 and 3.7 give the strength properties and uses of the main packaging papers and boards related to their basis weight. To reduce the weight of paperboard boxes and improve their strength, insulation and shock absorption characteristics, as for shipping containers, corrugated fibreboard is used. This type of board

Table 3.6 Main packaging papers

Basic material	Method of processing	Weight range g/m^2	Tensile strength kN/m	Properties and uses
Kraft papers	From sulphate	70–300	MD: 2.4–11.3	Heavy-duty paper, bleached, natural or coloured; may be wet-strengthened or made water-repellent; used for bags, multiply sacks and liners for corrugated board; bleached varieties for food packaging where strength required
Sulphite papers	Usually bleached and generally made from mixture of softwood and hardwood pulps	35–300	Very variable	Clean bright paper of excellent printing nature used for smaller bags, pouches, envelopes, waxed papers, labels and for foil laminating, etc.
Greaseproof papers	From heavily beaten pulp	70–150	MD: 1.7–4.4 CD: 0.85–2.1	Grease-resistant for baked goods, industrial parts protected by greases and fatty foods
Glassine	Similar to greaseproof but super-calendered	40–150	MD:1.4–5.2 CD: 0.85–2.8	Oil and grease-resistant, odour barrier for lining bags, boxes, etc. used for soaps, bandages and greasy foods
Vegetable parchment	Treatment of unsized paper with concentrated sulphuric acid	60–370	2.1–14.0	Non-toxic, high wet strength, grease – and oil-resistant for wet and greasy food e.g. butter, fats, fish, meat, etc.
Tissue	Lightweight paper from most pulps	20–50	Low strength	Lightweight, soft wrapping for silverware, jewellery, flowers, hosiery, etc.

[Source: (220)]

(see Figure 3.5) consists in a fluted-board layer 'sandwiched' between two sheets; two or three sandwich layers (the same or generally different) may be joined (Figure 3.5).

Corrugated cardboards can be used to replace wood for many purposes. The specific dimensions of various corrugated flutedboards are also indicated in Figure 3.5. The distance between two flutes is called the 'wave length'; the 'wave height' controls the board thickness. The manufacture of corrugated fibreboards and boxes began towards the end of the 19th century and grew rapidly in the early 20th century. Approval by the railways to replace wood for the shipping of many commodities was the key to the first growth of this industry. As mentioned before, a corrugated box is made from two or more sheets of liner board and one or more fluted sheets of corrugating medium. The liner sheets, i.e. the outer and the inner plies, are made from kraft and kraft-lined boards, while the corrugating medium is generally made from either beech or birch pulps or occasionally from straw and other waste materials. The corrugations or flutes vary and give the corrugated board different strength and cushioning properties. The table Figure 3.5 shows the properties of the various flutes. Generally speaking, C-flute and B-flute are the most common, but each type has its own special qualities: B-flute is less easily crushed, folds more easily and has stronger score lines than C-flute. C-flute has qualities that fall between A and B, making it a good compromise, and E-flute is a speciality grade, generally used for boards which are required to print very well.

Paper is stronger in what is called the machine direction than perpendicularly. This variation of the resistance to traction is not noticeable in the case of hand-made paper manufactured according to the dip process. The knowledge of the machine direction could be important in the design of packaging. Paper and board packages have the advantage of light weight; they are relatively clean and hygienic. Most paperboard packages can be stored flat: hence the empty packaging takes up much less space than glass, metal or rigid plastic packages.

The major disadvantages of paperboard packages are the tendency to absorb water, a high permeability to air and water

Table 3.7 Main packaging boards

Type of board	Construction	Characteristics	Principal uses
Chipboard 200–800g/m^2	100% low grade waste pulps from newsprint and fibreboards	Low cost, non printing, grey/brown, poor folders	Rigid boxes packaging pieces, fitments and tubes
Cream-lined chipboard 200–800g/m^2	Liner of mixed mechanical and chemical,pulps,backs chipboard	Lowest cost board which can be printed and folded	Lowest quality cartons
White lined chipboard 200–1000g/m^2	Liner is 100% new pulp mainly chemical, backs chipboard	Good printability, strength and foldability	Quality folding cartons for shirts, cosmetics, detergents, etc.
Duplex board 200–1000g/m^2	Liner and back both new pulp, middles mixture of both	As white-lined chipboard	ditto
Clay coated boards 200–600g/m^2	As white-lined with coated surface	Excellent printability	Top grade cartons
Solid bleached board (food board) 100-200g/m^2	100% virgin sulphate pulp	Strong board, high whiteness, odourless good folder	All foods, particularly frozen foods and baked foods
Liquid packaging boards 200–400g/m^2	As food board but with barrier coating or laminations (PE and Al)	As food board	Milk, creams, fruit juices etc
Kraft liner board 120–400g/m^2	More than 90% kraft pulp	High strength	outer component of solid board liners for corrugated board
Test liner 120-450g/m^2	Skin of kraft, rest of waste pulps	Burst strength equivalent to kraft liner grades	Liners for corrugated boards

[Source: Paine (220)]

vapour. Moreover, unless they are specially treated, paper and board are generally not resistant to liquids: under the influence of water, oil or other liquids their strength properties fall significantly. However, both paper and board can be 'upgraded' by additional treatments so that they can be used to package liquid products and decrease their permeability to gases and water vapour. Water resistance can be increased by the addition of resins, (e.g. urea formaldehyde resin) to the paper pulp. Most of the paper strength is due to hydrogen bonding between OH- and CO-groups on adjacent fibres. When water is removed during the paper-making process, these bonds are generated and the process is reversed on re-wetting. A wet-strength paper is one that remains strong when immersed in water. We must first of all distinguish between wet-strength papers, waterproof papers and moisture- or water vapour-resistant papers. The last-named must contain a barrier, usually a plastic film or metal foil to prevent the penetration of water vapour. A water-proof paper is one that sheds liquid water easily. Wet-strength papers, on the other hand, when they are saturated with water must still retain strength. By definition they should retain at least 30% of their dry strength when this property is measured by means of the burst strength test.

The penetration of the surface of any paper involves both the wetting of the surface and the mechanism of penetration through the paper capillaries. The packaging requirement must be considered before making a selection from the possibilities: do we require oil or grease resistance? is there a need for minimum staining? or do we want absolute resistance for a considerable period? Plastic-coated papers, parchment papers and so-called greaseproof papers, in that order, offer decreasing protection from oils, fats and greases. The degree of protection provided by the last three is dependent on the capillarity of the paper and the wettability of its surface.

Vegetable parchment paper was developed in the 1850s. It is produced in two stages: a pure chemical pulp is made into paper on the Fourdrinier machine and is then passed through a bath of sulphuric acid of controlled strength and temperature as well as the time of immersion. The parchmentizing process is stopped gradually by passing through diluted sulphuric acid and water, and finally the sheet is dried. This soaking in concentrated sulphuric acid (40%) swells the fibres tremendously and partially dissolves them. The swollen fibres close their pores, fill in voids in the network and produce extremely strong hydrogen bonding. When they are rinsed with water, this bonding is consolidated resulting in a paper that is oil-resistant, odour- and taste-free, and water-resistant.

Greaseproof paper or imitation parchment is made in the same way as ordinary paper, but the pulp is given an intensive beating action before going to the paper machine. It is a substitute for vegetable parchment and is much inferior in its properties.

Glassine is made by passing a greaseproof paper over a super calender, where the paper surface is given a very high finish and the density of the sheet is increased. This has a close-net structure and the hydrated condition of the cellulose enables the papers to be used for wrapping greasy products as well as products that are liable to lose flavour and aroma. Both greaseproof and glassine papers are frequently plasticized to further increase their toughness. They run well on high-speed packaging machines and can be chemically modified to enhance their wet strength. When waxed they are used for food pouches to package crisps, soups, cake mixes, etc. These upgrading methods, especially those involving the use of plastics (PE), make paper less capable of biodegrading. The cellulose paper components will in time vanish in the biological process during the formation of compost; but when they are coated with plastic, they will remain almost unchanged for a long time.

Many types of paper packaging are also unsuitable for recycling because of the plastic coating on both sides. Incineration is probably the most realistic method of using them for the recovery of energy, because paper and the protective plastic layer both have a high thermal content which is liberated during the combustion and can be recovered (heating buildings, steam production and hence the production of mechanical energy and electricity).

5 PLASTICS

a) Definition and historical introduction

Plastics are a group of long chain organic compounds of high molecular weight (30 000 to 10^6) formed by the chemical joining of smaller molecules together. They are produced either from high molecular elements existing in nature (e.g. cellulose, rubber, proteins) or are completely synthetic. At first, these latter 'completely synthetic' products were made from coal tar, however nowadays a wide variety of derivatives obtained from crude oil are used: either directly as the reactant (e.g. ethylene for polyethylene), or as an intermediate for the synthesis of the desired molecule (e.g. vinylchloride from acetylene for polyvinylchloride). Plastics are thus typical products from the petrochemical industry and constitute approximately 5% of the total crude oil consumption. The remaining 95% is used for electrical power generation, for heating and as motor fuels.

Although a wide range of basic compounds for the production of plastics were available around 1830, the development of high polymers and their use dates only from 1930. Marketing and the common use of plastics in packaging took place in the 1950s. Knowledge of the correlations between the molecular structure of plastics and their physical properties has enabled the synthesis of materials featuring the pre-required properties to be obtained by a wise choice of raw materials, reaction conditions, final processing and the use of some additives.

A survey of all the plastics available, of their structure and properties goes well beyond the scope of this chapter, but the principles on which these universally known products are based will be outlined here, to enable a better comprehension of the relevant environmental problems.

The properties of most synthetic plastics, apart from their insolubility in water and their progressive mouldability when heated, depend mainly upon their molecular weight; while the nature of the monomers and their method of combination will determine the properties relative to permeability and tensile strength of the final products.

The development of cellulose nitrate in 1845 to cope with the demand for a substitute for making ivory billiard balls, is often considered as the beginning of the 'plastic era'. In 1862 objects made of 'parkesine' were displayed at the London World Fair and 'celluloid' was patented by Hyatt in 1870: celluloid was produced by dissolving a mixture of cellulose nitrate and camphor, in a solvent of mixed alcohol and ether. The viscous concentrated celluloid solution is poured into a mould, and, after evaporation of the solvent, the material retains the mould shape. Celluloid was also made into film for cinematographic use. Cellulose acetate, soluble in acetone, quickly followed the very flammable celluloid and replaced it for film-making. Cellulose xanthate solution, from which regenerated cellulose film is produced by passing through a slot into a bath where it reacts with sulphuric acid, was developed in 1892. The first commercial film, known in the USA as Cellophane, and rayon fibres, both obtained by the viscose process, were introduced into Europe between 1912 and 1914, and in the USA in 1924. Regenerated cellulose film, now frequently coated with suitable materials, is still used as a packaging film, although its use is declining.

The range of completely synthetic plastics began with styrene, a liquid discovered in 1831 during the distilling of vegetable resins, and whose spontaneous polymerization under the influence of light was noticed. It was first produced in 1866 from benzene (present in coal tar), but polystyrene was not used for packaging in the food sector until the 1930s. Expanded polystyrene used for insulation and packaging purposes dates from 1950; at that time also Acrylonitrile-Butadiene-Styrene (ABS) was also developed after the discovery of the Buna-S rubber (Butadiene-Styrene) in 1933.

Vinyl chloride, the raw material for polyvinyl chloride (PVC), was prepared for the first time in 1835; the polymer was first advertised commercially in 1873 and was used for wire insulation in 1912. PVC shrinkfilm for packaging purposes appeared on the commercial scene as late as 1958. The discovery of vinylidene chloride (VdC) in 1838 has enabled, by the co-polymerizing it with PVC in the 1950s, to result in the production of 'Saran', which has much better properties (resistance to chemicals, insolubility in water and organic solvents) than PVC itself.

Polyethylene (PE) is by far the most common plastic used in the packaging sector (approx. 65%). Its manufacture from ethylene by Fawcett & Gibson of ICI, took place first in 1933: very high pressure was needed for its preparation, but by using so-called 'sandwich' catalysts containing titanium, K. Ziegler was able to prepare a polymer at low pressure during World War II. Conversely to the ICI product, the low pressure process gives a polymer with a higher density. Hence the difference between LDPE (low density PE) and HDPE (high density PE). In 1954 Natta discovered a stereospecific catalyst for the preparation of polypropylene (propafilm), a polymer derived from propylene that has taken over the greater part of the Cellophane market. The most recent developments concern mainly the composition of various plastics and aim to meet the desired protective characteristics for packages. Efforts have also been made to design 'degradable' materials to solve the so-called 'litter' problem. Biodegradable packages, heat-resistant 'cook in bag' materials enabling the heating and/or cooking of foodstuffs in their packages, other plastics designed for microwaves and ionizing radiation, potential multi-trip and/or more easily recyclable plastic packaging for sparkling soft drinks, are examples of what may exist in the near future. The development of the plastic market has certainly not yet stopped and there are plenty of potential new applications!

b) Some statistical data on plastics

The world production rates for the period 1900–1990 (Table 3.8) shows the development of plastics over the past 90 years. The steep increases during the 1930s and the 1950s could have been the consequence of the discovery of new products and processes for new purposes, such as the insulation of deep-sea telecommunication cables, and the need to cope with World War II. The increase in the 1960s surely resulted from the general trend towards pre-packaged products, parallel to the

Table 3.8 The world production of plastics (in 1000 tons) in the period 1900–1990

1900	20	1965	14 000	1980	49 000
1913	50	1967	17 990	1981	52 000
1924	60	1969	25 345	1982	48 000
1929	85	1970	28 186	1983	50 000
1939	350	1971	31 000	1984	53 000
1947	870	1972	37 000	1985	58 000
1950	1500	1973	40 000	1986	59 000
1953	2150	1974	41 000	1987	60 000
1955	2800	1975	35 000	1988	61 000
1957	4000	1977	43 000	1989	61 000
1960	6900	1979	51 700	1990	62 000

[Source: *Plastica* 32, 9 (1979) *Plastica* 36, 1 (1983) and APME (187)]

trend of the expansion of the self-service shops during the same period. The remarkable rise in the use of plastics can thus be explained as follows:

1. innovation and new applications (insulation for electric machinery developments, and the automobile industry, etc.),
2. penetration into the markets of more traditional materials such as metal, cement, leather and paper, and the extension of those markets into the building trades, the furniture business, and the textile and leather industries,
3. developments and common usage in packaging.

Production rates indicate that we are now at the beginning of a more stable, mature period which means that innovation is going to be the main lead to noticeable new rises or changes. Production and consumption data per country, expressed as a percentage of the total, are given in Table 3.9. These also show that actual consumption (production plus import minus export) per capita varies in Western societies from 30 kg to 180 kg per capita.

Generally the consumption of plastics tailored to packaging reaches 23% to 33%. The building trades (26%), the packaging sector and the electronic industry together claim approxi-

Table 3.9 Plastic production expressed as % of the total world production and plastic consumption rate (per capita) in different countries over the period 1977-1990

Country	Production in % of the total world consumption (a)	Annual consumption per capita(in kg)* 1977 (a)	1980 (b)	1988 (c)	1990 (c)	Production per country 1000 tons 1989 (d)	%WE (d)
USA	34.5	64	65.7	99.0	104	27,000	–
West Germany	14.6	89	73.1	130.9	137	6,000	23.9
Japan	13.6	41	58.8	89.6	92.6	11,000	–
Russia	7.2	13	15.1	–	–	–	–
France	6.2	48	50	64.0	68.0	3,611	13.9
Italy	5.8	41	45.1	70.0	74.0	4,000	15.4
Great Britain	5.7	44	51	60.5	61.1	3,4454	13.3
Spain	2.3	32	31.9	54.4	57.9	2,053	7.9
Others, namely	10.1						
Belgium	1.4	44	48.2	144.0	151.0	1,424	5.5
The Netherlands	1.1	31	34.5	75.0	–	1,120	4.3
Sweden	–	79	106.4	–	–	822	3.2
Austria	–	71	71.4	101.7	117.9	766	3.0
Switzerland	–	60	39.1	93.2	96.5	627	2.4
Finland	–	92	123	83.0	81.0	423	1.6
Norway	–	57	99.2	73.8	–	310	1.2
Denmark	–	83.4	97.2	92.7	100	470	1.8
TOTAL EEC	–	–	–	71.3	–	22.947	88.6
TOTAL WE	–	–	–	73.3	–	25.905	–

* Concerns only the plastic consumption in the national plastic processing industry
[Source: (a) *Plastica,* 32 (1979), 9 p 272; (b) *Plastica,* 236 (1983), 1, p 26–31; (c) IPAD (1992); (d) *Kunststof en Rubber* (1992), 10, p 36]

Table 3.10a Consumption rates (in 1000 tons) of various plastic resins according to their application in Belgium

Year/plastic	Phenol resins	Polyester	PE	PS	PVC	Cellulosics	Others	Total
1980	19	14.6	169.5	51	100	88.4	45.8	488.9
1981	24.7	22.5	170.5	52.9	97.4	101	43.9	512.9
1987	46.4	33.3	261.9	61.2	126.6	125.3	41.9	696.6
1988	36.4	33.1	272.8	69.0	143.4	137.2	57.7	749.6
1989	28.3	37.0	311.5	66.3	166.9	143.9	74.5	828.4

Sector/categories	1980	%	1981	%	1987	%	1988	%	1989	%
Packaging	118.8	26.9	112	23.5	155.4	23.3	173.8	23.1	192.1	22.8
Sheet and film	96.2	21.8	107.7	22.7	208.1	31.2	226.5	30.1	264.5	31.4
Household*	24.4	5.5	27.5	5.8	48.6	7.3	54.4	7.2	51.1	6.1
Building	137.4	31.0	125.6	26.4	164.8	24.7	193.4	25.8	227.2	26.9
Sport	1.2	0.3	1.6	0.3	1.0	0.2	1.2	0.2	1.0	0.1
Others	64.0	14.5	100.8	21.2	88.9	13.3	101.9	13.6	106.7	12.7

* Commodities, furniture, electronics, etc.
PS: Consumption of plastics in the automobile industry for Belgium and Luxemburg: 199 000 tons (1989)
[Sources: National Institute for Statistics, APME]

Table 3.10b Consumption rates (in 1000 tons) of various plastic resins according to their application in the Netherlands

Year/Plastic	LDPE	LLDPE	HDPE	PS	PVC	PP	ABS	Other	Total		
1980	190	–	65	100	180	65		150	750		
1987	162	20	83	117	204	109	7.5	–	702.5		
1988	165	26.5	87	126	231	138	7.6	–	778.1		
1989	168	29	90	143	230	169	9.2	–	838.2		
1990	190	27	102	144	223	184	9.0	–	879.0		
Sectors/categories									Total (1980)	%	Total % 1989
Packaging	135	–	15	40	25	10	–	–	225	30	33
Building	10	–	10	20	100	10	–	40	190	25	20
Commodities	20	–	30	25	30	25	–	60	190	25	10
Others	25	–	10	15	25	20	–	50	175	20	25
Agriculture	–	–	–	–	–	–	–	–	–	–	5
Transportation	–	–	–	–	–	–	–	–	–	–	7

PS: Consumption of plastics in the automobile industry for the Netherlands: 37 000 tons (1989)
[Sources: *Plastica* 36 (1983), 1; *Kunststof en Rubber* 12 (1990) and 10 (1991)]

Table 3.10c Plastic resins consumption (%) in Belgium and the Netherlands 1980-1990

Plastics	Belgium (%) 1980	1989	The Netherlands (%) 1980	1989	1990
LDPE	55.0		60.0	20.0	21.6
LLDPE		37.6		3.5	3.1
HDPE	10.0		6.0	10.7	11.6
PVC	15.0	20.1	11.0	27.4	25.4
PS	10.0	80.0	17.0	17.1	16.4
PP	7.0	–	4.0	20.2	20.9
ABS	–	–	–	1.1	1.0
Polyester	–	4.5	–	–	–
Phenolic res	–	3.4	–	–	–
Cellulose	–	17.4	–	–	–
Others	3.0	9.0	2.0	–	–
Total	100.0	100.0	100.0	100.0	100.0

Table 3.11 World consumption rates (in 1000 tons) of the five main plastic resins

Country	Plastic resins PVC	LDPE	HDPE	PP	PS+EPS
USA	4082	5428	4010	3738	2297
Japan	2028	1621	958	1793	1277
W. Europe	4745	4982	2807	3369	1611
Canada	418	511	281	186	188
Others	2188	1416	1081	1176	963
World total	13 467	13 958	9137	10 312	6270
in 1990 %	25.33	26.27	17.19	19.47	11.80
World total	10 225	10 350	4860	4315	6358
in 1980 as %	28.3	28.7	13.5	12.0	17.6

[Source: Kato, H, 'Trends of the General-purpose Plastic Resin Industry in 1982'; *Packaging Review Japan*, 3 (1982), No 9 p 26–34]

mately 55%–65% of the total plastic market (Table 3.10). Typical consumption rates of the five main plastic resins are given in Table 3.11. Although the consumption rate for PVC has become more important, the packaging sector still consumes far less PVC than PE. The average consumption rates for different plastic materials in Belgium and the Netherlands over the period 1980–1990 are given in Table 3.10.

c) Production of some main plastic resins and their environmental impact

As already mentioned, a large range of plastics is available nowadays. In order to understand the environmental problems created by the industrial production of plastics, it is necessary to keep in mind the broad lines of every chemical process. The products are generally produced in a liquid medium, by heating or cooling in a reaction vessel. Usually the plastic product is not the only result of the reaction; other compounds are produced. This means that, after separation, the product must be purified and that some (unwanted) substances remain in the reaction medium. The BOD-quotient and the quantity of suspended matter in the effluent from the reaction medium are too high to allow untreated discharge into the sea, lakes or rivers. A major problem is the presence of small quantities of some specific (toxic) pollutants, such as traces of catalysts, solvents, monomers, malodorous products (containing sulphur or nitrogen) which are difficult to eliminate.

An additional problem can arise from the pH-regulation (neutralization) of the aqueous medium: as the organic molecules react slowly, it could happen after a while that new bonds are formed, even after being passed through the waste water treatment, so that the traditional and automatically controlled systems do not ensure the purity required for waste water disposal. Thus the plastic industry has a major problem of waste water treatment (see RD dated 3 August 1976).

Plastics resins used in the packaging sector can be divided into two major groups according to the raw materials used to produce them.

Semi-synthetic resins

The semi-synthetic resins in which natural high molecular compounds are chemically 'modified' (i.e. not broken down into smaller molecules). Cellulose (see paper and board) is typically an important natural product for the synthesis of packaging films of Cellophane, cellulose acetate, ethyl cellulose and cellulose nitrate. Regenerated cellulose film is produced by dissolving pure cellulose fibres in an alkaline solution of sodium hydroxide and carbon disulphide, where the compound sodium cellulose xanthate is produced. This soluble cellulosic derivative is a pink-coloured (sulphuric bond), syrupy solution. When forced through a long slit into a bath of diluted sulphuric acid, the cellulose is regenerated as a clear, transparent film. For packaging use it must be treated with plasticizers such as glycerine or ethylene glycol. The plasticized film cannot be heat-sealed, and is still very permeable to water vapour. Therefore it is coated on one or on both sides with thin films of nitrocellulose and waxes, or completely synthetic plastics such as PVDC. Thus, there are many different cellulose film products according to thickness and coating. The reaction scheme and the commercial codes for cellulose-films are given below:

$$(C_6H_{10}O_5)_n + nNaOH + nCS_2 \longrightarrow$$

$$(C_6H_9O_4\text{–O–C}{\genfrac{}{}{0pt}{}{\nearrow S}{\searrow SNa}})_n + nH_2O$$

Cellulose 3 free OH group per glucose monomer	sodium hydroxide	carbon disulphide	⟶	cellulose-xanthate soluble in conc. NaOH solution

Cellulose-xanthate reacts with sulphuric acid to form the insoluble cellulose (but now as film) CS_2, water and sodium sulphate (Na_2SO_4). Plain and treated Cellophane are coded as follows:

PT:	plain transparent cellulose film
QMS:	semi moisture-proof cellulose film (one side coated)
MSAT:	nitrocellulose-coated cellulose film (moisture-proof)
MXXT/S:	copolymer solvent coated cellulose-sealable
MXXT/A:	copolymer dispersion coated cellulose-sealable
PE/MSAT:	PE laminate on one and nitro-coat on other side of cellulose film

Because of the many sulphur bonds, generally malodorous and toxic (e.g. the volatile carbon disulphide), which escape into the atmosphere and pass into the effluents during the regeneration process, the industry producing cellulose film also has to take some measures to limit its typical and dangerous odour nuisance (CS_2 and derivatives).

Cellulose films have in the past commonly been used to package bread, cakes, gingerbread, toffees, chocolate snacks, cookies, potato chips, etc. Good results were also obtained for the wrapping of pharmaceuticals and textiles needing protection against moisture, by moisture-resistant cellulose film. Because of the environmental problems of odour and effluent, as well as cost, a large part of the cellulose film market has been taken over in the 1980s by polypropylene film.

Cellulose acetate is formed by the reaction of cellulose with acetic anhydride; it is soluble in dimethylketone and acetone. The volatile solvent is then evaporated from a 30% solution of cellulose acetate and is recuperated in a loop when films or fibres are made. The very clear films are used in laminates and as 'windows' in paperboard cartons and boxes.

Acetate film is almost insoluble in water but features a good permeability to oxygen and carbon dioxide gases as well as to water vapour, so that it is adequate to package goods such as fruits, vegetables and flowers, which 'breathe'.

Completely synthetic plastics
The completely synthetic plastics are obtained partly from crude oil derivatives (ethylene, propylene, benzene, toluene, etc.) obtained by cracking naphtha and partly from second generation products such as styrene, vinyl chloride, ethylene glycol and phthalic acid.

The petrochemical industry is based on crude oil products mainly by cracking to produce a range of hydrocarbon fuels and lubricating oils. Both oil refineries and petrochemical industries have a number of waste disposal problems. Those of the petrochemical industry are the most difficult because of the large variety of intermediate and final products. A range of water pollutants are the main concern; the gaseous substances are generally combustible and can be burnt before escaping to the atmosphere. Water consumption and effluent production have now decreased in the petrochemical industry leading to higher concentration of undesirable components and the waste water treatment technology must take this into account. Although nowadays only an average of 1 to 3 m^3 of waste water is produced per ton of ethylene, the pollution potential remains substantial due to present production rates (>1000 ton/day). The water consumption (and thus the waste water output) of a petrochemical plant is divided (according to German statistical data) as follows:

85% cooling water (potential thermal pollution)
6% boiler feed water (potential but very limited atmospheric pollution)
8% 'process' water (water from reactor vessels, hence potential water pollution)
1% sanitary water

Untreated effluents contain some inorganic and organic, soluble and insoluble substances.

Table 3.12 gives average values, based on data from 25 petrochemical plants, for some of the parameters mentioned in Chapter II. The same table indicates also the standard that discharge water must meet according to Belgian regulations. The range of both pH-values and suspended matter is typical and very large in the petrochemical sector.

Table 3.12 Characteristics of effluents from 25 petrochemical plants compared with Belgian Governmental requirements

Parameter				Norms: discharge into superficial water		
	Min- (a)	Max (b)	Average (1)	General (2)	Petrochemical industries (3)	Oil refineries (4)
Temp (°C)	30	35	–	30	35	30
pH	5	9.2	–	6.5<pH<9	6<pH<9	6.5<pH<9
Suspended solids (mg/l)	50	200	80	60	300	60
Sedimentable solids (mg/l)	–	–	–	0.5	1	0.5
Hydrocarbons (mg/l)	5	100	35	5	25	20
Fats & oils (mg/l)	3	30	10	10	–	25
Detergents (mg/l)	1	10	3	3	–	3
BOD (mg/l)	20	500	150	15	300	50
COD (mg/l)	200	2000	600	–	2000	300
Total nitrogen (mg/l)	–	–	–	–	–	40
Total phenol (mg/l)	–	–	–	–	–	1

[Source: (1) Cadron (course notes); (2) RD 3/8/1976; (3) RD 22/4/1977; (4) RD 3/8/1976]

Completely synthetic plastics can be classified according to the reaction principle involved:

1. *Polycondensates:* these start with a single monomer or a binary mixture of various monomers, each having at least two reactive groups. During the successive reactions it is always the latest compound formed that reacts (molecular growth up to high molecular weight) every time with the elimination of water. The general mechanism is as follows:

A + B —> A–B + D
A–B + A —> A–B–A + D
A–B–A + B —> A–B–A–B + D etc.
A–B–A–B is the plastic (polycondensate);
abbreviation –[A–B]n–.

A typical example is polyester, one of the main plastics for packaging, polyethylene terephthalate (PETP) produced from phthalic acid and ethylene glycol by condensation in autoclaves at high pressure and temperature in the presence of catalysts.

n phthalic acid + n ethylene glycol ⟶ PETP + n H_2O

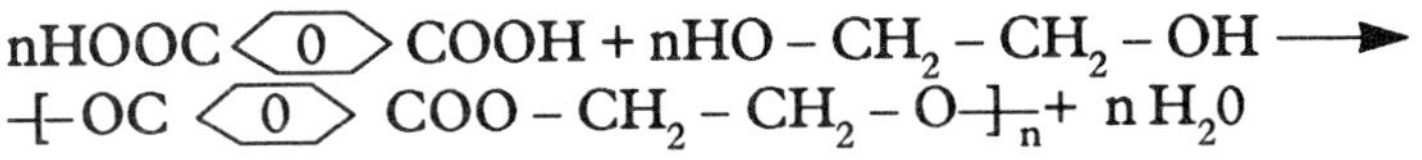

In the USA this polyester is known under the trade name (Du Pont) 'Mylar'. The ICI product is called 'Melinex'. Due to its good mechanical properties, PETP is designed to package solid (hard or sharp) foodstuffs, e.g. bags for frozen products. Nowadays PETP bottles are also used to package soft drinks: they resist the overpressure created by the carbonic acid gas for which they provide a good barrier.

2. *Polymer compounds:* The difference between the reactions leading to the formation of this type of plastic and those relative to polycondensates is the fact that no small molecules, such as water, are liberated. The reactivity of the monomers in question lies in the 'double' bonds. Under the influence of special catalysts (peroxides such as dibenzoyl peroxide) and favourable pressure and temperature conditions, the 'double' bonds open creating two reactive sides per molecule; the resulting polymerization reaction can be represented as follows:

$$n\,CH_2 = CH_2 \xrightarrow[+\text{ catalysts}]{\text{high temp \& pressure}} n\ -[CH_2-CH_2]- \longrightarrow -[CH_2-CH_2]-n$$

ethylene molecule (monomer)	activated condition during the polymerization process	polymer PE

High-density polyethylene (HDPE; d = 0.940–0.965) withstands temperatures up to 125°C and can therefore be sterilized by boiling water and is used for containers (e.g. bottles for milk and boxes) as well as for films. Low-density polyethylene (LDPE; d=0.910–0.925) is generally used as film and as a coating to 'upgrade' paper and board. LDPE has a melting point lower than HDPE, cannot be sterilized and is less suited for the packaging of oils, fats and oxygen sensitive products.

Polypropylene is another important 'polyolefin'. Chemically similar to PE, as shown by its formula:

monomer $CH_2{=}CH{-}CH_3$– polypropylene (PP) $-[CH_2-\underset{\displaystyle CH_3}{\underset{|}{CH}}]-_n$

it has even better physical properties and lower permeability to oxygen and water vapour. Sacks woven with polypropylene strips serve to package vegetables, fruits and grains.

If one of several other atoms (e.g. chlorine, fluorine) or groups (e.g. hydroxyl, acetyl, phenyl) replace one or more of the hydrogen atoms in the olefine, a range of polymers (the vinyls) is obtained which feature quite different properties. In use for the packaging sector are:

polyvinyl chloride (PVC)

$$-\!\!\left[CH_2-CH(Cl)\right]\!\!-_n \text{; monomer: } CH_2=CH(Cl)$$

polyvinylidene chloride (PVDC):

$$-\!\!\left[CH_2-C(Cl)_2\right]\!\!-_n \text{; monomer: } CH_2=C(Cl)_2$$

polyvinyl acetate (PVA):

$$-\!\!\left[CH_2-CH(OOCCH_3)\right]\!\!-_n \text{; monomer: } CH_2=CH(OOCCH_3)$$

polystyrene (PS):

$$-\!\!\left[CH_2-CH(C_6H_5)\right]\!\!-_n \text{; monomer: } CH_2=CH(C_6H_5)$$

polymethyl methacrylate (PMMA):

$$-\!\!\left[CH_2-C(CH_3)(COOCH_3)\right]\!\!-_n \text{; monomer: } CH_2=C(CH_3)(COOCH_3)$$

The properties of these 'vinyl' resins depends mainly on the substitution atom or group. They are generally more crystalline (thus harder) than polyolefins and their permeability to gases is lower. They generally have better resistance against fats and oils. Mineral waters and oils are packaged in PVC bottles; solid fats (margarines) and yogurts in PS boxes. Meats and cheeses are also wrapped in PVC film. The copolymer of VC and VdC, known as Saran, serves as a coating for other films to improve the gas and vapour barrier characteristics. For these reasons, packing gaskets and other packaging accessories are made of vinyl resins. Other main plastics for packaging purposes are:

Fluorinated polymers Plastics (e.g. teflon) very resistant to acids (used as gaskets), having low permeability to gases and water vapour. Because of their high cost, they are mainly used as packaging accessories only when other plastics would not be adequate.

Ionomers Plastics where 'ionic' bonds between the polymer chains are realized by insertion of metallic ions (e.g. zinc); hence the resistance properties are improved. An ionomer similar to PE is marketed under the name 'Surlyn A'. It is more resistant than LDPE to high temperatures and can be applied in thinner 'coatings'. Because of their good tensile strength, ionomers can also be used to package frozen products.

Epoxy resins These possess remarkable adherence to metal and good chemical resistance; hence they are used as lacquers to protect metal cans.

Ethylene vinylacetate copolymer (EVA) A plastic copolymer produced from a mixture of ethylene and vinyl acetate. As a stretch film, EVA serves to wrap meat (e.g. chicken) and, like shrink film, a tight wrapping is possible by tension, but without heating.

d) Some properties of plastic resins determined by the molecular structure

Thermoplastic properties

The plastics we have described briefly are longchain molecules. If polymers are made from monomers having three reactive groups per molecule (e.g. glycerine with 3 OH groups), we would obtain three-dimensional plastics. Long chain linear molecules are 'thermoplastic': they soften when temperature is raised because the 'Van der Waals' forces which hold them together are weakened. The three-dimensional so-called 'duroplasts' (thermosets) do not soften or melt with temperature increase, but will breakdown when heated to about 350°C. Most plastics used for packaging are thermoplastic. This characteristic is useful to produce the various packaging forms. To avoid breakdown and/or oxidation of these organic materials during the heating and forming process, stabilizers and antioxidants are added. Moreover, additives will improve weather resistance (protection against the influence of atmospheric oxygen, of rain and of UV light) – see also Table 6.7.

Barrier properties

In practice the structure of the different polymers is not as simple as described above: not only does the length of the polymer chain vary, but there are other structural discrepancies (e.g. lateral branching) which hinders the close packing arrangement of the molecules. Thus, inside a plastic sheet there will be some areas with an ordered (crystalline) structure and some with a disordered (amorphous) structure of the polymer molecules (see Figure 3.6). In the disordered amorphous areas the greater intramolecular distance leads to a higher permeability to gases.

If for any reason (e.g. specific reaction conditions or stereospecificity of the monomer as encountered in styrene, vinyl chloride and vinyl dichloride), the structure of the plastic molecule (e.g. styrene, PVC and PVDC) is ordered, the degree

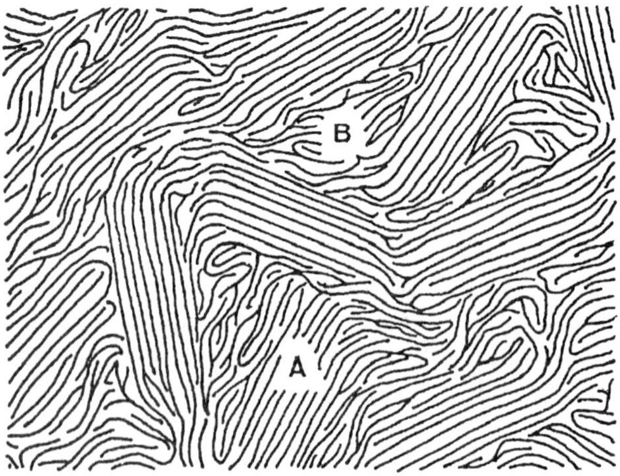

Figure 3.6 Diagram illustrating crystalline (A) and amorphous (B) areas in plastic

of crystallinity of the material rises (and its resistance to tensile stress also increases) and the permeability to gases decreases. The tension exerted on films induces a better arrangement of the molecules when stretched, hence the tensile strength and the barrier properties can be improved by mechanical treatment (orientation).

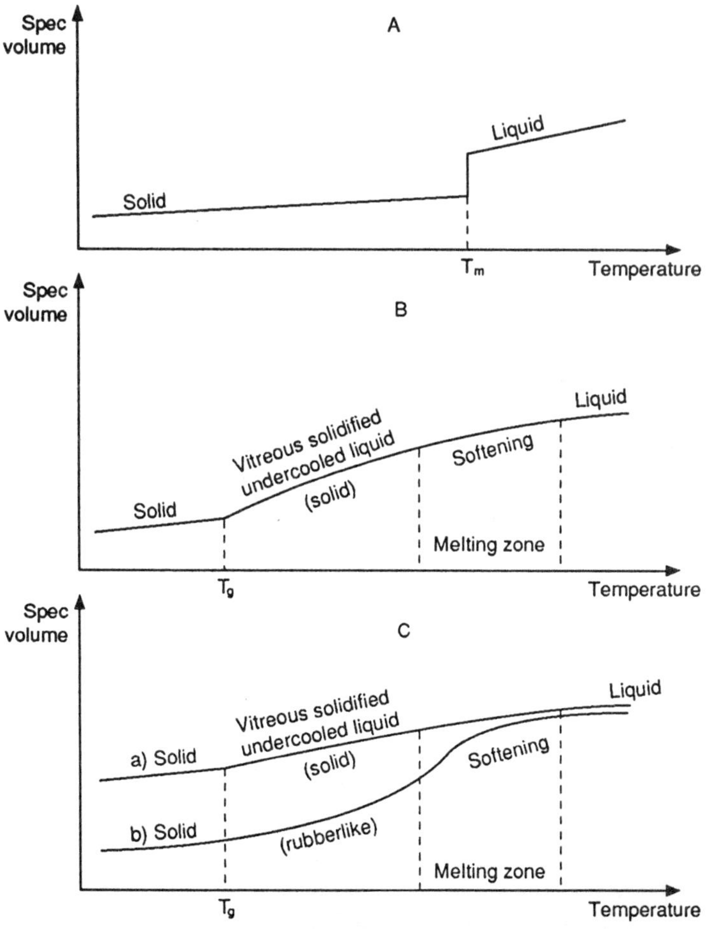

Figure 3.7 Illustration of physical behaviour of plastics at melting point and at glass-transition temperature

Softening or glass transition temperature 'Tg'

The transition temperature Tg, like the melting point, is a major parameter determining the utilization properties of a polymer. It indicates the point where a refrigerated liquid reaches the vitreous state. On a diagram giving the specific volume as a function of temperature, there is a bending point at the glass transition temperature. Figure 3.7 illustrates various possibilities:

1. For a crystalline substance, there is a clear change at the melting point T_m (transition from solid to liquid and vice versa).
2. For a glassy substance (amorphous structure) there is no precise melting point but an intermediate state where the material (e.g. glass) softens enough to be processed. Cooling it will then reset to form a highly viscous liquid, hence the vitreous state is perceived as a solid.
3. As mentioned earlier, plastics have both crystalline and amorphous areas. For instance, in HDPE their respective share amounts to 80% and 20%. Plastics feature the specific properties of both crystalline and amorphous substances. Whereas for glass the Tg is much higher than room temperature, for many plastics its value ranges around the usual ambient temperature (Table 3.13). When the Tg of a particular plastic is somewhat higher than the room temperature (20°C), it is in a rather semi-crystalline state (hard, brittle, fragile, less flexible, etc.)

Should the ambient temperature be higher than Tg, the plastic would behave like a chilled highly viscous liquid (vitreous state) or like rubber. Its behaviour depends on the molecular mass and on the kind of molecules. In that state the material is flexible and can be easily used in automatic packaging machinery.

Shrink films are plastic materials of a high molecular mass and a Tg markedly higher than the ambient temperature.

To manufacture them, suitable films are heated to above the transition temperature, when they are then stretched, elastically in (one direction or in two directions perpendicular to one

Table 3.13 Comparison of various plastics with respect to their temperature dependent behaviour

Class	Polymer	Glass transition temperature (Tg) in °C	Melting temperature in °C	Decomposition temperature in °C
Poly-olefins	HDPE	–123	135	360
	LDPE	–60	110	335
	PP	–10	170	330
	TPX (poly-4-methyl-pentene)	30	235	335
	Ionomer (PE-Na acrylate)	–110	120	340
Vinyls	PS	82	240	325
	PVA	71	240	250
	PVC	80	(210)	160
	PVDC	–20	190	200
	(SARAN)	–17	150	150
	PTFE	130	330	475
	PMMA (plexiplars)	65	160	170
Ethers	POM (polymethylene oxide)	–85	(181)	100
	NC (tri-niitrocellulose)	60	(700)	160
	Cellulose	230	(450)	270
	CA (tri-acetylcellulose)	105	(306)	250
Ester	PET	81	265	283
	PC	147	267	330
Amides	Nylon 6	40	(220)	200
	Nylon 66	50	(265)	200

[Source: Oswin (125)]

another), and subsequently cooled under tension; these stretched films will 'shrink' back to their initial dimensions when they are reheated briefly above Tg. Hence the great variety in applications: tight packaging of meat, cheese etc., or bundling of a number of unit packages such as cans, bottles, and boxes (palletized or not).

The specific properties, namely glass-transition temperature, elasticity, melting point, permeability, etc., depend very much on the molecular structure. Most packaging purposes require plastics (films) with the highest crystalline structure (low permeability to gases), having also a good flexibility and elasticity (Tg < room temperature). These opposing requirements can be met only by the addition of some substances to improve the properties (e.g. plasticizers).

e) From plastic material to packaging

Plastic resins are generally marketed in the form of granules or beads. Before being used, the material is mixed with some additives, namely antioxidants, stabilizers, plasticizers etc.,

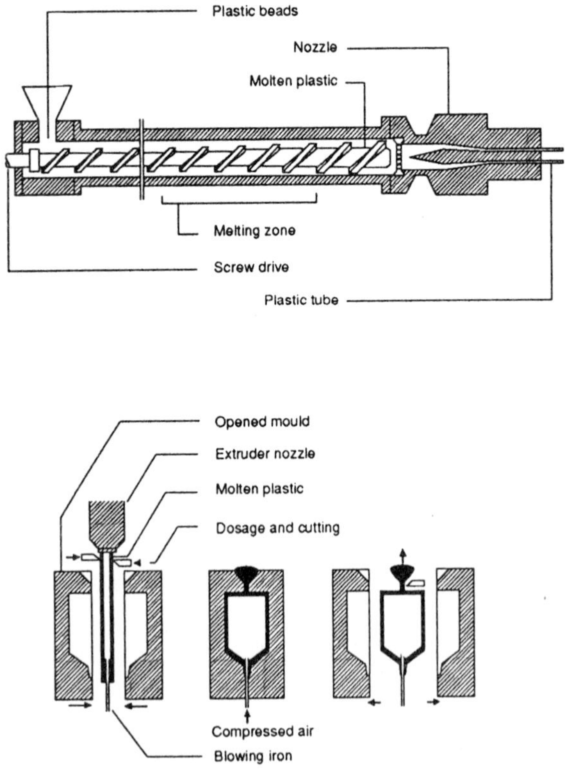

Figure 3.8 Sketch of an extruder and bottle-shaping

according to the purpose (foodstuff packaging, transport packaging, shrink film etc.) and the required characteristics.

Usually packages are formed directly from the molten plastic mass, either by forcing it into a closed mould (injection moulding) or by 'blowing' a hollow tube against the internal surfaces of a mould by means of a blowing iron connected to compressed air (blow moulding). The latter process is similar to the shaping of glass containers (see Figures 3.1 and 3.8).

Many simple packages, namely blisters, cups, bags as well as complex ones (laminates) are made of thin plastic sheets. Sheets and films can be produced continuously by forcing the

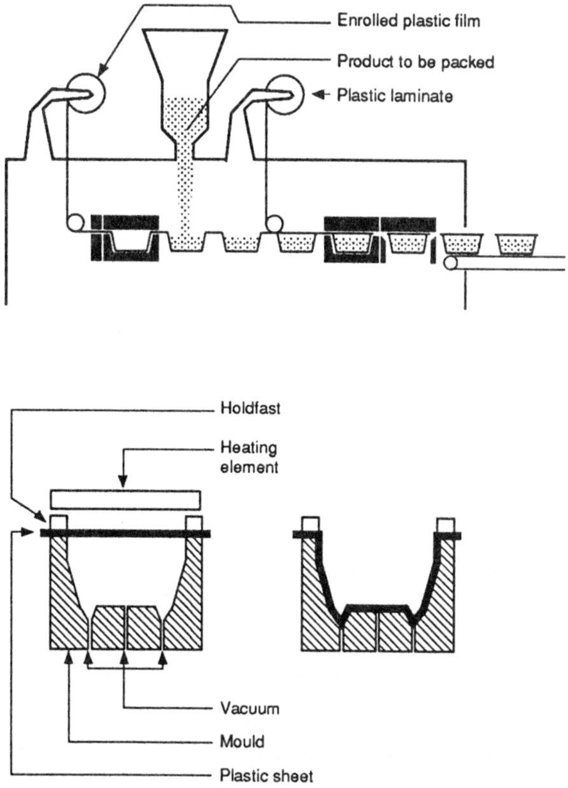

Figure 3.9 Sketch of an integrated packaging machine

molten plastic mass through a long narrow slit (extrusion). Very wide plastic sheets in PE are made by extruding a seamless tube from a circular die; during cooling, the tube is expanded by compressed air and finally the tube is slit longitudinally and opened out.

It is thus essential to melt the plastic – in the 'extruder', the plastic beads are gradually heated by means of electric heating elements, and a slow rotating Archimedean screw forces the plastic mass through the 'nozzle' (see Figure 3.8). In the case of thermoplastic resins, welding and sealing are achieved by hot compression. Figure 3.9 is a sketch of an integrated packaging machine: it illustrates how unit packages are obtained in a continuous mode: first the softened plastic film is shaped into a cup, then, after filling, it is closed with a second (laminate) by heat sealing. These operations require only electric power.

f) Recycling of plastics

Taking into account the thermoplastic properties of most plastic resins for packaging, we might suppose at first sight that homogeneous post-consumer plastic packaging could be returned to the extruder for remelting and reshaping (regeneration after regranulation). However the material must meet certain requirements for cleanliness and homogeneity. Studies of plastic materials and their processing, have shown that successive thermal treatments lead to the formation of some decomposition products and that to some degree oxidation (breakdown) reactions will occur. Thus the material of the final plastic container is not the same as the virgin material from which it was made. This explains also why plastic packages are generally not designed to be reused. The difference in molecular structure and polymer properties increases in direct proportion to the time between manufacture and potential recycling. With some plastics, 'fresh' waste (e.g. scrap collected at the extruder) can well be regranulated and mixed to some extent (normally about 20%) with virgin material.

As plastic resins vary considerably from one another in their nature, structure, additives, composition (copolymers, laminates), use, etc., it seems impossible to process a mixture of plastic resins recovered from household wastes to produce products that must meet specific requirements. It is perhaps feasible to mix homogeneous plastic refuse, namely PE, PS, PVC or PETP, with adequate 'additives' to manufacture lower grade products (e.g. flower pot holders, rubbish bags); however, up to now the cost of collection, sorting and cleaning has exceeded the price of new materials. An actual general policy to reduce the amount of refuse to be disposed of and to restrict the spoilage of materials, prompts the recovery and revaluation of wastes. Separate collection of post-consumer plastic packaging can be arguably justified: to increase the efficiency of separation and hence the retreatments, plastic wares should carry a label identifying the resinous class to which they belong, or they should be the object of a return system by the imposing of a deposit. The incineration of plastic mixtures present in household refuse is nevertheless from the economic point of view, the most realistic treatment (see also the heat content of plastics, Chapter IV).

Some homogeneous and mixed plastic wastes can also be profitably treated by 'pyrolysis' (heating in the absence of atmospheric oxygen). This method was in fact applied for the first time by Lavoisier, to assess the chemical composition of organic products and afterwards used on an industrial scale for the production of coal gas.

In the case where thermoplastics (e.g. PE, PS, PP) are treated by pyrolysis, an efficient 'thermodegradation' to the monomers from which they were prepared can be achieved. Of course, the products obtained in this way may be reprocessed to chemicals or to the polymers themselves (true recycling). The whole process requires energy and has seemed up to now to be justified only for some special cases, as for the treatment of used tyres. Pyrolysis of products containing the element chlorine (e.g. PVC, PVDC) will result in the release of hydrochloric acid and a series of new products, but monomers are not formed. Recycling PVC thus means the recovery of hydrochloric acid, from which chlorine can be produced (neutralization and subsequent electrolysis). At least the chlorine can be used as a reagent with ethylene to form again the monomer vinyl chloride.

g) Degradable plastics

We have already referred to the problems 'spoiling' the beauty of Nature, of recreation areas or streets with packaging carelessly abandoned or dumped ('litter'). While paper packaging wastes may disintegrate by exposure to water and the elements and may be almost completely eliminated in a relatively short time by biodegradation, plastics do not undergo such processes because most plastics are water repellent, hence potential enzymatic reaction becomes difficult. Furthermore, those biological enzymatic reactions which can break down high molecular weight products such as proteins, starch and cellulose, always start at the ends of the molecules; the probability that in a plastic item some molecules end just at the surface is very low; moreover the structure of most plastic molecules does not lend itself to the desired breakdown.

We know organic molecules can degrade under the influence of sunlight (UV), oxygen and water; however this process too is significantly slowed down in the case of plastics due to the presence of necessary antioxidants. A PE bottle left on a beach will become friable only after two or three years of exposure to the elements and only then become ready for further breakdown. Burying or controlled dumping of plastics limits the natural effects of exposure to the elements; and here they are preserved even longer (10 years or more). However, some successful attempts to reduce the 'life-time' of plastics already exist. It is possible to insert into the polymer chain, spots which are highly 'sensitive' to light from the sun (or to UV light during waste treatment) and to atmospheric influences – and at these places the molecules will be attacked, will break into smaller pieces that become more soluble in water and thus liable to biodegradation. Also, light-sensitive antioxidants can be selected which lose their protective properties faster, enabling reactions to start more rapidly (173, 174).

In the last decade, better results in making plastics more apt to biodegradation have been achieved by adding biodegradable products such as starch or chemically modified starch to the polymer matrix. The reduction of the polymeric material hereby obtained for the production of the so-called 'biodegradable plastics' also contributes to their accelerated disappearance in nature; but the starch, being enveloped with the synthetic material such as PE, will not degrade by biological means unless access has been created by degrading the surrounding polymer. So exposure to light to induce photodegradation and oxidation remains necessary for these types of plastics.

A series of real biodegradable plastics was produced by ICI and came on the market in 1990 under the trade name 'BIOPOL'. Principally, the polymer is a polycondensate of hydroxybutyric acid and variable amounts of hydroxyvaleric acid. The hydroxy-acids are produced and bound to a homopolymer (polyhydroxybutyrate) or to a copolymer (where a mixture of hydroxy-acids is produced) dependant on the substrate and various conditions by which the selected bacteria are forced to grow.

BIOPOL has a linear structure and is thus a thermoformable plastic: some characteristics and properties are collected in the table below:

Copolymer properties	0%	10%	20%
Melting point (°C)	180	140	130
(Tensile strength (MPa)	40	25	20
Young's modulus (GPa)	3.5	1.2	0.8
Extension to break (%)	8	20	50

[Source: ICI (224)]

As in the cases of the natural high molecular products such as cellulose, starch, proteins, BIOPOL will disappear biologically when buried by means of the bacterial fauna present in the soil, and does not need any exposure to 'induce' degradation.

The production of the 'so-called' and the 'real' biodegradable plastics may not be regarded as an attempt to stimulate littering but as a method of solving some practical problems of recovery and disposal of plastics encountered in the agricultural sector where a large amount of films are used on the soil for a well determined period of the year, with the function of improving crop and fruit production.

Besides the fact that biodegradable plastics can be functional for use in some cases (medicine, agriculture), their importance for large scale applications (such as in the packaging sector) is relatively small. First of all their cost is high, but they are in fact useless for food conservation in the long-term. Furthermore, because of their predisposal to decay naturally by composting processes, they belong unquestionably to the category of the throwaway or single-use materials and, when disposed, have no real advantage especially when material and/or energy recovery are taken into consideration.

We may also conclude that the presence of degradable and/or starch containing polymers is likely to be undesirable when plastics are recovered with the intention of reprocessing.

IV
Packaging and energy

1 INTRODUCTION

It is generally agreed that a high standard of living and a high energy consumption are linked (27). Of the parameters of the civilized development level, the industrial and transport sectors together represent about 70% of the total consumption of energy (see Table 4.1). This has been confirmed in particular by the relevant data from West Germany and Sweden which in 1976 reached respectively 77% (5) and 60% (6).

Table 4.1 Comparison of average energy consumption in agricultural and industrialized communities (in 10^6 Joules per day and per capita)

Sector	A Agricultural community	as % of the total	B Industrialized-community	as % of the total	B/A
Food	25.0	23	42	4.4	1.7
Household and trade	50.2	46	276	28.7	5.5
Agriculture and industry	29.3	27	380	39.5	13.0
Transport	4.2	4	263	27.4	63

[Source: Lox (111) from *Scientific American*]

Since publication of the 'Club de Rome' report and the energy price increase after the 1973 oil crisis, people have become aware that raw materials and energy resources should be used efficiently. Each consumer is concerned because higher energy costs at the production stage are, of course, finally included in the market price of the goods. Packaging is an expression of style and standard of living; it is also an industrial product requiring different kinds of energy.

In this chapter, we intend to compare energy consumption during the successive steps in the lifetime of different packaging alternatives. Such a study is important because the use of energy can generate either heat or atmospheric pollution. Moreover, it gives a picture of the price structure and shows where potential savings lie.

The examination of the energy consumption needed for glass packaging (cf. Table 2.1 Chapter II, p 77) indicates that conversion of raw materials consumes the most energy. A similar statement is valid for the other packaging materials. Hence, most saving can occur when the quantity of material required per item is reduced (material and energy savings), or when a better technology is applied (energy saving), or, if necessary, by selecting another material or another resource of energy (price and/or ecological considerations). Such a survey has been conducted by the Swedish Packaging Institute which found that costs are lower if unbleached instead of bleached paper pulp is used for milk packaging, not only because the process requires less energy, but also because less materials are needed (103, 6). To be fully comprehensive, we must note that the energy needed in the manufacture and distribution of all the packaging used represents for each country in Western Europe only a small part of their total energy consumption. Lidgren (103) found its share in 1976 was approximately 2.4% for Sweden, 2.7% for Belgium and 1.1% for Europe (estimation based on references 27 and 103), without taking into account the recoverable energy present in used packaging. The sectorial survey remains important for the industrial sectors concerned; moreover, the data indicates ways to select a feasible waste disposal treatment.

The energy analyses mentioned in the literature can be misinterpreted and are likely to give rise to controversy because:

1. The different energy sources needed to produce and manufacture packaging materials are averaged over the field as a whole, yet the cost of each source of energy and/or raw

material may vary considerably from one country to another, e.g. Sweden prefers to use electricity generated from hydraulic power and wood because these are cheaper than oil and petrochemicals.

2. The ratio of electric power required to primary energy cost, depends on the production process and therefore is not the same in all circumstances and/or countries. So, the result is quite different when converting water power (watermills, turbines driven by water in dam facilities, or tidal power) into electricity from that using energy derived from the calorific power of burning oil, or coal, or heat generated from nuclear reactions to produce electricity, which have an efficiency of only 25% to 30%. Moreover, the impact on the environment is not the same: in particular, hydro-electric power plants, in opposition to thermic power plants, are non-polluting.

3. In some calculations, the combustion heat (quantity of energy generated by the combustion of household refuse in incinerators) should be deducted from the total since this source of energy has been recovered.

In Chapter V, we discuss the fact that the caloric value of household refuse cannot be considered as constant. Consumers often burn combustible materials such as paper, board and plastics; there are other ways to treat and/or to recover them, e.g. the recycling of paper or the pyrolysis of plastics. Note also that, since industrial development, people have steadily paid little attention to the recovery of materials and energy even though it would have been feasible. Little importance has too often been given to the direct or indirect effects on the environment.

2 ENERGY PRODUCTION AND ENVIRONMENTAL FACTORS

Two forms of energy, mechanical power and thermal energy, are used in the manufacture of packaging materials, while mechanical power is used mainly in transport and distribution.

The raw materials from which these forms of energy are produced can be considered as combustibles, because the mechanical power required for motion and traction is generated – either directly in the internal combustion engine or indirectly via electricity production – by means of the heat produced by the combustion of fuels. While the two energy forms can be converted into one another, the efficiency is not the same for both cases. Mechanical power can be converted almost 100% into heat (Joule Effect) whereas the conversion of heat into mechanical power follows the second law of thermodynamics, with a much lower yield. The average is about 25–30%, depending on the temperature limits within which the heat can be supplied and taken out, hence there is a tremendous release of heat to the environment (70–75%) responsible for thermal pollution. Fuels can be divided into two categories, namely the so-called primary and secondary fuels: the first are used in their original form, secondary fuels are obtained by processing of primaries. Figure 4.1 outlines the reciprocal relationship between raw materials and forms of energy.

Electricity production requires mechanical power to drive the generators which produce electricity. In the so-called thermal power plants the generators are connected to the steam turbines. Motors with internal combustion can also be used to drive the generators, but this applies to power plants with reduced capacity. In hydro-electric power plants, the generators are driven by turbines which convert the potential energy of the water held behind the barrages, or the kinetic energy of tide or rivers, directly into mechanical power.

The conversion efficiency of mechanical energy into electric power is very high (ca 85%). Electricity can be transported at high voltages with a high yield; losses are quite limited when the electric power is converted into heat (heating elements) or converted into mechanical power (electric motors). As electricity can then be converted again into one of the two forms of energy (heat, mechanical energy), it can also be considered as a secondary fuel. Primary fuels on the one hand and secondary fuels on the other hand, are for the most part interdependent:

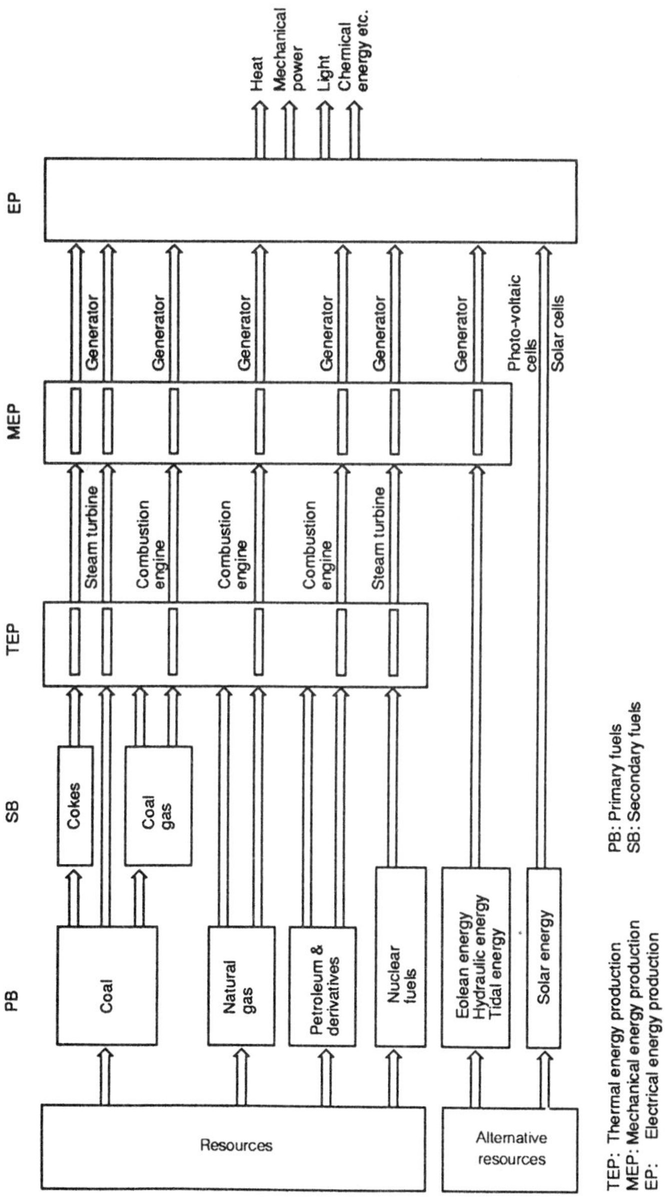

Figure 4.1 Diagram showing the relationship between primary and secondary forms of energy

e.g. the extraction of coal (primary fuel) requires electricity (secondary source) for lighting, ventilation, air compression, the driving of elevators and conveyors, etc.

Steam engines (heat produced by the combustion of coal) or diesel engines, can provide the tractive force needed to transport coal or crude oil. Hence, plants producing fuels inevitably consume a part of what they produce. Likewise the fractional distillation process to separate petroleum into its components (gas, petrol, diesel oil, paraffin oil, fuel oil, etc.) consumes fuel oil. This all means that the production of fuels and power requires energy and releases heat (thermal pollution). The power used to transport fuels is generally considered as negligible, e.g. the quantity of fuel consumed to transport oil by sea is negligible in relation to the tanker capacity!

The energy production efficiencies are determined by the ratio between the energy content of the delivered form of energy and the total consumption of all involved energy sources. The Joule ($kg.m^2/sec^2$) is the unit of energy. Caloric value or thermal content is defined as the total amount of heat thermal content produced by the complete combustion of a fuel; the thermal content unit is also the Joule. The unit of electric power used hereafter in the comparison tables is the 'kWh' (kilowatt-hour); since 1 Watt is equal to 1 Joule per second, 1 Watt-hour = 3600 Joule, and 1 kWh equals one thousand times as much, $3.6.10^6$ Joule.

Table 4.2 gives comparative data on the energy needed for production (extraction, refining, transport), the energy content of a unit quantity and the efficiency of the process. It shows clearly that coal and natural gas production efficiencies are high, but that the yield relative to the production of electricity from coal is very low; hence, the production of electricity from heavy fuel and diesel oil could be even more unfavourable considering the energetic aspects and the possible air pollution by the combustion gases. Of course, the efficiency is higher if the intermediate step of heat to mechanical power is bypassed. The yield relative to the production of electricity from waterfalls in hydro-electric power plants is approximately 85% which means that the potential and/or kinetic energy of water is converted

Table 4.2 Energy content, energy consumed in the production of various energy sources and the efficiency of production

Energy resource	Quantity considered	Energy consumed in production (MJ)	Energy content (MJ)	Total energy required (MJ)	Production efficiency (in %)
Coal	1 kg	1.39	28.01	29.40	95.0
Coke	1 kg	3.93	25.42	29.35	86.6
Electricity	1 kWh	10.80	3.60	14.40	25.0
Natural gas	1 m^3	2.61	31.70	34.35	92.4
Heavy fuel	1 litre	8.50	40.98	49.55	82.7
Light fuel	1 litre	8.29	40.18	48.47	82.9
Kerosene	1 litre	6.96	36.53	43.49	84.0
Diesel oil	1 litre	7.42	37.83	45.25	83.6
Petrol	1 litre	6.85	35.97	42.82	84.0
LPG propane	1 kg	8.89	50.00	58.89	84.9
LPG butane	1 kg	8.89	49.30	58.19	84.7
Lubricating oil	1 litre	8.29	40.18	48.47	82.9
Wood (dry)	1 kg	–	19.75	–	–
Rubber	1 kg	–	27.50	–	–
Wood (water content 12%)	1 kg	8.00	17.20	25.20	68.3

[Source: Boustead and Hancock (22); Ahlhaus (5b)]

into electricity with a 85% efficiency. In this case, heat is not necessary, so there is no thermal pollution. Such considerations lead to the conclusion that the impact on the environment depends on the way thermal, mechanical and electric power are produced.

When producing mechanical power from heat, approximately 70–75% of the heat must be dissipated. In electric power plants, this loss can cause thermal, surface water or atmospheric pollution; in the first case, the river water is pumped into a loop to act as a cooling medium; in the last case, the heat is dissipated in cooling towers and passes into the atmosphere in the form of water vapour. When burning fuels to produce steam heat, or when fuels are used for internal combustion engines (direct conversion into mechanical power), an amount of heat is dissipated in the atmosphere through the chimney or expelled with the exhaust gases (ca 54% in the case of the heat generation process and ca 79% in the case of transport). Exhaust gases mostly contain air polluting components (see Table 4.3). Estimated emission rates of some energy consuming sectors in the USA are given in Table 4.3. From this, it seems that the sectors of transport, heating and industrial processing (solid waste incineration included) are nearly equally involved in air pollution.

Table 4.3 Estimated emission rates of air pollutants in the USA (in 10^6 ton per year)

Sector	CO	Solid particles	SO_2	HC	NO_x	Total
Transport	63.8	1.2	0.8	16.6	8.1	90.5
Heating (combustion)	1.9	8.9	24.4	0.7	10.0	45.9
Incineration of solid wastes	7.8	1.1	0.1	1.6	0.6	11.2
Industrial processing	26.6	17.1	7.9	13.1	1.9	66.6

CO: carbon monoxide; SO_2: sulphur dioxide; HC: hydrocarbons (incomplete combustion of fuels), NO_x: mixture of nitrogen oxides

For nuclear power plants, where the heat gain comes from the mass deficit Δm due to the fission of heavy atom nuclei, e.g. uranium (cf. the Einstein equation $E = \Delta m.c^2$), thermal pollution will also occur for the same reasons as stated for the conventional power plants. The environmental problems also include:

1. the possible dispersion of radioactive elements in the air and in the water if leakages occur in the nuclear reactor,
2. the possible release of radioactive materials due to the chemical treatment of nuclear fuel. This process also produces

radioactive refuse which is now partly stored in tanks, partly sunk into the sea. Leakages result in a dual pollution – the wastes contain the toxic heavy metals, which are at the same time radioactive,
3. the decommissioning of nuclear power plants (estimated lifetime 30 years): pollution of buildings housing the nuclear reactor and the adjacent fuel reprocessors by radioactive materials is almost inevitable!

If we do succeed in perfecting nuclear fusion (here, the mass deficit caused by the union of light hydrogen nuclei is converted into heat) to produce energy, the present nuclear refuse problem will be mainly solved.

3 ENERGY CONSUMPTION FOR PACKAGING MANUFACTURE

The total amount of energy required for the production of packaging materials and the other forming and manufacturing operations that lead to the final package, can be expressed per packaging unit or per unit of weight. We have chosen the comparison per material on a weight basis because this physical unit is more representative when studying simple and combination types of packaging. Consider, for example, the closure of a glass bottle: nowadays we use an aluminium threaded cap with a plastic wad. Each part can be characterized by its weight. The same quantity of material can be consumed to manufacture several packages, the content and weight of which may vary; it is obvious that a big glass bottle is heavier than a small one, but bottles of the same capacity can also vary in weight. If we wish to give a particular shape and/or outline to a type of package for special reasons, mainly psychological, more material might be required than would be necessary to manufacture a purely functional model; glass packages for cosmetics usually weigh more than ones for beer or milk. Moreover, returnable packages are mostly designed to be thicker and thus heavier – to limit breakage – than single-trip throwaway packages of the same volume. The available data also demonstrate that the energy necessary to transform packaging materials into a package is generally in direct proportion with the weight of the materials: it means that the manufacture of 18 bottles of 250 ml capacity weighing 173 g each with 2112 g hot glass mass, consumes almost the same energy as the forming of three bottles each of 1 litre weighing 704 g each. We will establish hereafter that the fuel consumption for the transport of materials depends much more on the weight than on the volume with a similar loading capacity.

Where possible, a distinction has been made by the nature of fuel to give a better understanding of the potential environmental effects. This analysis can partly explain the cost of packaging; however, it cannot take into account the labour costs relative to its handling. Often the analysis for a particular material appears more favourable than for another which competes for the desired use, but the final choice, based on economic parameters, often turns to the less favourable packaging from

Table 4.4 Comparison between weight and capacity of some throw-away glass bottles designed to package non-alcoholic drinks

Capacity (in ml)	Weight (in gram)	Capacity (in ml)	Weight (in gram)
129	162	600	440
190	223	700	489
193	159	750	496
193	281	977	630
199	173	1000	546
250	176	1000	567
250	188	1000	616
250	195	1000	670
250	208	1000	740
250	216	1014	465
293	249	1018	670
334	380	1136	703
341	196	1136	710
500	347	1176	740

[Source: Boustead, Hancock (22)]

the energy viewpoint. This is valid if the labour costs are higher than the raw materials costs. As already shown in Table 2.1, returnable glass packaging is more favourable from the energy point of view than equivalent throw-away packaging; nevertheless the latter is preferred nowadays because the labour costs borne by the distributor for collection, distribution, storage and transport are reduced. If these tasks were automated, at low cost, returnable glass would probably become more widely used.

In the meantime, selective collection of glass through bottle banks may be considered a good method of tackling energy and waste problems (see Chapter V). The survey of the energetic aspects of packaging includes also the determination of their energy content. The caloric value and thus the energy that could be recovered in the form of heat by burning them is indicated in the tables under the heading 'energy content'. Unlike most published tables, we do not include here the fuel consumption for the transport of raw materials and/or final packaging forms, because the structure of the traffic and the distances vary from one country to another, and can be calculated on estimates for each separate case.

a) Glass packaging

During the last 30 years, automation in the production of glass containers and a better control of the composition of glass have enabled the manufacture of thinner packages of strength at least equivalent to that of the old thick ones. The weights of some throwaway glass packages for non-alcoholic drinks, as a function of their content, are indicated in Table 4.4.

One litre glass bottles for milk weigh approximately 700 g; beer bottles containing 340 ml weigh approximately 200 g. The energy required by the raw materials and glass production is given in Table 4.5.

Table 4.5 Energy requirements (in 10^6 Joules) to produce 1 ton of packaging in glass and metal

Material	Production process	Electricity	Petroleum products	Others	Energy content	Total
Glass	Raw materials	80.000	2.276	942.000	–	3.298
	Glass	1.224	6.195	4.977	–	12.396
	Total glass	1.304	8.471	5.919	–	15.694
Tin can	Iron production	369.000	2.433	13.210	–	16.012
(170 ml)	Tin plating	1.365	4.507	2.895	–	8.767
3TP	Package manufacture	2.104	5.495	11.700	2.015	19.299
	Cover and bottom plates	6.165	7.761	5.010	3.469	18.936
	Total tin can	10.003	20.196	32.815	5.484	63.104
Tin can	Iron production	1.516	5.953	14.921		22.390
(450 ml)	delivery	6.000	179.000	–	–	185.000
3 TFS	Package manufacture	1.806	4.119	10.018	1.927	15.943
	Cover and bottom plate	7.203	6.511	4.054	2.962	17.808
	Total tin can	10.531	16.802	28.993	4.889	56.326
Aluminium	Production of aluminium	53.316	42.668	13.370	21.567	109.354
(450 ml)	Production Al-plate	3.601	9.534	9.905	–	23.040
	Package manufacture	5.892	9.334	24.537	2.230	39.763
	Cover-closure (easy opening)	16.816	15.293	9.464	6.913	41.573
	Total Al-can	79.625	76.829	57.276	30.710	213.730

3TP: three piece can; TFS: tin-free steel
[Source: Boustead, Hancock (22): recalculated values from data]

The consumption is almost the same to produce coloured or clear glass. Small differences have their origin not in glass-making itself but in the production of the raw materials giving colour to the glass. As we have discussed before, glass production involves raw materials which are mixed with 20–25% broken glass ('cullet'); this is necessary to facilitate the melting process at lower temperature. On average, 80% of the necessary cullet is supplied by the glassworks itself from its own production rejects. The other 20% comes from glass removed from household refuse or collected separately. Increasing the proportion of cullet leads to lower energy consumption during glass-making: bringing the supply up to 40% would enable a 25% energy saving, but the risk of reducing the quality at higher recycle rates (with fluctuations in quality also possible) must be considered.

It is useful to examine the average life-time of returnable glass packages: it can be calculated on the basis of a certain loss coefficient. For example, let us suppose losses during the journey between filler and consumer and back amounts to *a*%, hence for each cycle of 100 packages (100 – *a*) bottles will come back; on the other hand, per cycle and per hundred *a* new bottles are introduced into service. Thus each package will be replaced by a new one after on average 100/*a* = n journeys there and back. If we consider, for instance, a population of glass bottles for milk, among them there will be *a*% new ones; however, the main part includes bottles which have been used at least 1 to n times. The average life-time, expressed as a number of cycles, can be worked out as follows (162):

$$L = \frac{\{1 + 2 + 3 + \ldots + n\}}{n}$$

The sum 1 + 2 + 3 + . . . + n is equal to n(n+1)/2, hence:

$$L = \frac{\{n(n+1)\}}{2n} = \frac{\{n+1\}}{2}$$

Practically, a batch of bottles with 3% loss coefficient covers on average 100/3 = 33 times the cycle from the filling company to the consumer and return. The average life-time, calculated on the total number of bottles the company issues, is at any moment (33 + 1)/2 = 17 uses. Taking these data into account, it appears in the example mentioned in Table 2.1 (bottles for beer the average life-time of which is eight uses) that the loss coefficient amounts realistically to 6%. In the case of returnable packages, the energy consumed for the transportation from consumer to the filling company should not be added: the consumer is supposed to go back to the shop for further purchases;

the additional consumption due to the presence of empty glass packages in a personal car is negligible. The same argument applies to the transport of empty bottles between the distributor and the filling company: a lorry delivering its load is always supposed to come back; in that case, the additional consumption due to the presence of empty glass packages is around 10% maximum. Finally, in the total energy spending, this part remains very low (cf. Table 2.1). That 'extra consumption' can in most cases be justified, especially when energy savings during raw materials exploitation and glass-making are taken into account. Some debaters (15) will insist on the potential environment spoilage induced by the washing (utilization of detergents) and the rinsing of the returned bottles. These objections should not be accepted because, in most cases, every bottle intended as a container for foodstuffs, medicines and chemicals is washed and rinsed before filling for hygienic reasons.

These considerations show clearly that the energy consumed per glass packaging decreases in direct proportion with the increase of the number of returns: if the return frequency is 50 (a = 1%), the saving in energy is approximately 50% (15, 103). Previous arguments have made it clear that the energy required for the delivery of glass bottles can be subdivided into two parts:

1. An amount Ei which is return-independent. Ei is an amount which is necessarily consumed whether the bottle is re-used or not: each bottle (new or re-used) must be washed, filled, closed, sterilized, labelled, handled etc.
2. An amount E_1 which is return-dependent: instead of producing x new bottles, only one is sufficient to organize one usage and (x–1) re-uses. So the energy to produce material and packages can be spread over the number of uses.

If we consider the energy-requirements for both cases, the non-return and return system, the following equations can be stated:

(a) production of x new bottles: $E_{1'tot} = x(Ei + Er)$
(b) production of one bottle and x uses ([x–1] returns or re-uses); $E_{2tot} = xEi + \frac{Er}{x}$

Hence, the energy gain amounts (see Figure 4.2) are :

$$E_{1tot} - E_{2tot} = Er\left(x - \frac{1}{x}\right) = Er.x \text{ when } x >> 1.$$

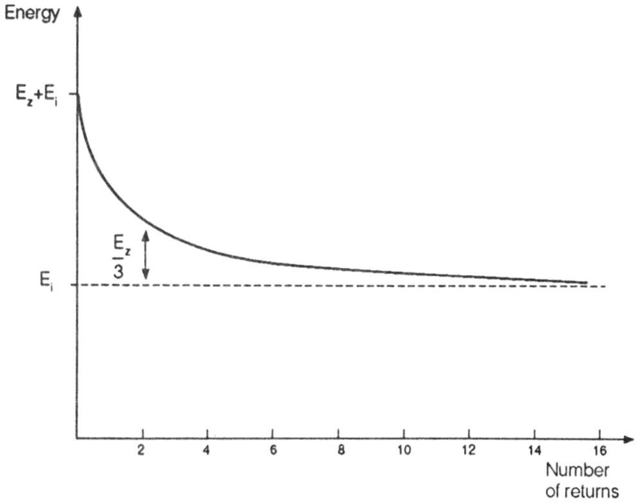

Figure 4.2 Diagram showing the evolution of the energy required per unit of packaging (e.g. glass bottles) as a function of the return rates

It has been calculated (Bojkow) that return systems become interesting from the energy point of view when x equals at least 10 (breakage rate: 5%). Therefore, we may also assume that situations become more interesting when Er > Ei, as observed in the case of glass.

For plastics we have an inverse situation: Er < Ei; assuming that plastic quality will improve in the near future to make returnables possible, we will still have to cope with the quality decrease of the material as a function of time (degradation by weathering) radiation, although the package is not breakable. Whether the material will resist and fulfill the quality requirements after ten returns or more, remains up to now an unanswered question. Throwaway packaging undoubtedly involves material and energy squandering if dumped; as mentioned earlier, the costs in materials, energy and especially in manual handlings, are of major importance for the current options in glass processing.

b) Metal packaging

Blackplate, tinplate and aluminium are the few metals involved in the manufacture of metal packaging. Aluminium cans are mainly used for carbonated beverages; Al-foil for wrappings and laminates. Steel is also used in the soft drink sector; however, it is used more in the packaging of foodstuffs (e.g. vegetables, fruits, cooked soups, condensed milk, oils), lubricating oils, solvents (ether, chloroform, hexane etc.).

Table 4.6 gives indications of the material needed (weight) for some metal boxes. Metal boxes consist of three pieces: a cylindrical (sometimes rectangular) body, an upper plate and a bottom plate. Aluminium boxes are generally manufactured in two pieces; the cylindrical body and the bottom are formed at the same time. After filling, this piece is closed by an upper cover (see Figure 3.2).

In the soft drink sector, all types of metal packaging receive an internal protective epoxy coating. This organic material adds to the energy content of the package; as this amount is added as material, but not consumed as energy, it is not included in the total energy consumption (cf. Table 4.5). Although solder is no longer used for making cans, the weight of which has diminished over the last years, these data are still interesting to demonstrate differences and possible improvements in the future. The available data also show that the energy needed for manufacturing the package depends mainly on the material, so the quantity per packaging unit can be worked out from the weight and the data from Table 4.6. Aluminium production from bauxite as a raw material mainly requires electric power (production by electrolysis). The quantities needed to generate the electric current and to produce the

Table 4.6 Amount of components (in grams) for metal containers as a function of the volume they represent

Volume	170 ml	285 ml	340 ml				455 ml			3.5L
Components	3TP	3TP	3TP	3TFS	2TP	2A1	3TP	3TFS	2A1	3TP
Body										
Tinplate	20.50	25.80	30.30	–	31.70	–	39.20	–	–	239.00
Tin-free steel	–	–	–	30.10	–	–	–	37.50	–	–
Aluminium	–	–	–	–	–	15.80	–	–	17.60	–
Cover										
Tinplate	–	–	–	–	–	–	–	–	–	72.60
Aluminium	3.70	5.05	5.05	5.05	5.05	5.05	5.05	5.05	5.05	–
Bottom										
Tinplate	7.30	10.50	10.50	–	–	–	10.50	–	–	72.60
Tin-free steel	–	–	–	10.20	–	–	–	10.20	–	–
Coatings										
Varnish	0.48	0.48	0.53	0.68	0.62	0.31	0.64	0.82	0.35	2.16
Ink	0.15	0.07	0.08	0.08	0.08	0.06	0.11	0.11	0.08	0.90
Sealings										
Solder	0.37Sn	0.64Pb	0.43Sn	–	–	–	1.04	–	–	1.10Sn
Glues	–	–	–	0.13	–	–	–	0.18	–	–
Closures										
Cover plate	0.05	0.09	0.09	0.09	0.09	0.06	0.09	0.09	0.06	0.15
Bottom plate	0.05	0.07	0.07	0.07	–	–	0.07	0.07	–	0.15
Total (gram)	32.60	42.70	47.90	46.40	37.50	21.30	56.70	54.00	23.10	388.70

3TP: three-piece can; 2TP: two-piece can; 2A1: two-piece can in aluminium; 3TFS: three-piece can but of tin-free steel; Sn: tin; Pb: lead
[Source: data according to Boustead, Hancock (22)]

oil products can be estimated on the basis shown in Table 4.2, relative to the yield of fuels. Although the weight of aluminium packaging is approximately half that of similar ones in iron, it appears that the energy needed is twice as much; thus a package made of tinplate is over twice as favourable to an aluminium box because it requires approximately four times less electric power.

Due to their construction, metal packages are impractical as returnables. Their scrap value is higher than for glass; moreover they are recyclable. Their magnetic properties enable an easy separation of the iron-containing materials. In the case of aluminium, its low density is used as a criterion for its separation from wastes. Tinned wares can be chemically treated in order to recuperate the valuable tin by electrolysis. 980 kg iron and 5.2 kg tin are in this way obtained from 1000 kg tinned iron scraps. This operation requires 75.06 kWh electricity and ca 30 litres of fuel oil as combustible. Iron and aluminium scraps are sent back to the melting furnaces which leads mainly to material recuperation but not to energy recuperation. Of course, heat could be recovered from these packages by the combustion of the organic components but this share is not really important. The scrap value of tin-free steel is less. The proportion of recycled iron in the production of cans is approximately 21.6% (22).

Among all of the metals used to manufacture packages, aluminium is of course the easiest to recycle by re-melting. Losses are at a maximum (ca 11%) when the magnesium percentage is high. 'Easy-open ends' of aluminium containers for soft drinks are made of an alloy containing more magnesium than that of the cylindrical body; therefore they are sometimes treated apart. From 1.1 kg scraps, we obtain on average an aluminium bar (ingot) of 1 kg. A total energy of 30.10^6 Joule is needed for this reprocessing (22). As approximately $109.5.10^6$ Joule are required for the production of 1 kg aluminium from raw materials (cf. Table 4.5), this recycling involves a 70% saving in energy.

The average proportion of these aluminium bars processed with freshly separated metal to manufacture new packages reaches 27%. This has been taken into account to calculate the data of Table 4.5.

c) Paper and board packaging

As discussed in Chapter III, there is a very wide variety of papers and boards; the differences arise mainly from the nature and the quantity of binding agents and resins used during processing. The base material remains cellulose obtained mostly from wood, present in the form of fibres accounting for about 50% of the weight of the cellulose resource. However, the binding agents do not greatly influence the energy consumed by paper-making, as the main part (ca 85%) is needed to separate the wood fibres from the wood and to dry the paper pulp. For this reason, an average consumption rate is given in Table 4.7. The energy needed to manufacture packages from paper and cardboard sheets is negligible compared with that consumed during the production process (ca 3.4%).

The feasibility of paper recycling has been studied in detail in Chapter III.

The energy consumed per package can also be worked out, taking the weight into consideration, from the data mentioned in Table 4.7; so, the average weight of one litre milk cartons, made of light board (ca 200 g/m^2) with double plastic coating, is 26 g (21 g board + 5 g PE for waterproof coating) (15).

Although paper recycling does not use energy profitably, paper can, in contrast to glass and metal recycling, be considered as a valuable 'combustible' because of its energy content. Its combustion liberates a quantity of heat (cf. Table 4.7) available for the generation of mechanical or electrical power. If we want to incinerate household refuse, the presence of paper is almost a must to ensure the combustibility of the whole.

Finally, we would like to point out that no toxic product is generated by the complete combustion of paper: as shown by Figure 2.1, relative to Nature's cycle, oxidation (reaction with oxygen) of vegetable material (wood) liberates water and carbon dioxide, two products stored as the 'solid state' (cellulose) in plants, thanks to the solar energy. The bleaching of paper pulp by chlorine can induce the production of dioxins, which are released to the atmosphere when the combustion is not performed in optimal conditions. However, there is a trend to replace chlorine by oxygen for bleaching paper.

d) Plastic packaging

Although many plastic resins do exist, only a few are used in the packaging sector; the share of polyethylene (PE) reaches approximately 65%. As already mentioned, crude oil is the main raw material for plastics; they are typical products of the petrochemical industry where crude oil and derivatives are used as feed-stock and as a source of thermal energy. Plastics have thus a heat content similar to that of the raw materials from which they are made; hence, their high calorific value. Overall, we consider that the production of plastics requires as much energy as that stored in the final material. It means that approximately 2 kg crude oil are needed to produce 1 kg plastic material (66, 67).

To make the final package, first the plastic is heated to near its melting point, then the appropriate shape is obtained by pressing. As illustrated by the data in Table 4.7, these automatic machines run exclusively on electric power (cf. Figure 3.9). The sequence of operations is totally identical to the manufacture of PE or glass bottles; the noticeable differences between these two types of packaging are the working temperature for shaping (ca 1000°C for glass, ca 150°C for PE) and the weight per unit (ca 700 g for glass, ca 20 g for PE).

Table 4.7 also mentions the energy consumed to process the plastic resin PETP (polyethylene terephthalate) used to package soft drinks. 'PET bottles' have a capacity of 1.5 litres or 2 litres and were originally provided with a base made of PE. Bottle and base weighed respectively 60 g and 5 g for 1.5 litre bottles and 65 g and 15 g for 2 litre bottles. For comparison purposes, we have also mentioned the consumption rate per ton of final product from which the share per package can be worked out. Studies on re-use or recycling of plastics are still in progress; in the food sector, recycling of plastic packaging

Table 4.7 The energy required in MJ to produce 1 ton of paper, board and plastics

Material	Production process	Electricity	Petroleum	Other	Energy content	Total production
Wood (1000 kg)	Logging	–	6640	–	17 200	6640
Paper pulp	Wood→chips	3060	–	16 300	–	19 360
Paper (1000 kg)	Total product wood→paper	6410	19 670	16 630	17 890	42 710
Board (1000 kg)	Total product wood→board	9350	25 630	16 630	17 890	51 620
LDPE	Petroleum → Gasoil → Ethylene → PE	2760	36 820	–	49 950	39 580
	Production of bottle from PE (50 000 bottles)	3960	–	–	49,950	43 540
HDPE	Production of bottle from petroleum (50 000 bottles)	6890	37 910	–	50 850	44 800
PP	Petroleum → Gasoil → Propylene → PP	3340	40 390	–	52 650	43 730
PET	Production PET	5390	71 180	–	46 560	76 570
	Production 16 670 bottles (60g/bottles 15 litre)	18 660	–		46 560	95 230

[Source: Calculated from data, Boustead, Hancock (22)]
Note: 1 ton oil equivalent (TOE) = 43.2 MJ

materials is forbidden in Belgium by the RD dated 24.08.1976. In principle, plastics, like glass, of the same nature can be re-melted and given a new shape; Chapter III explains the reasons why this is not achieved in practice for good quality food packaging. Of course, plastic packaging waste can serve to manufacture products designed for quite different purposes: the manufacture of flower pots, rubbish bags, etc., and can be recycled in the form of monomers or recovered for their energy content. In fact, the 'recycling' to monomers is expensive in terms of energy and is in practice applied only to process rubber refuse. Thermosets can be re-used when ground to powder as a filling material to be mixed with virgin plastics before moulding.

In Chapter V, we shall discuss the amount of plastics which are recovered nowadays from such different sources as waste consumer packaging and household rubbish, and from buildings, cars and agriculture (PE-films).

e)Packaging accessories

Energy data related to packaging accessories, such as paper or plastic inserts, labels and pallets should also be considered. Due to their purpose, closing systems, for instance, are designed for a single use (cf. Chapter I) and we may wonder whether the energy consumed is important compared with that needed for the total package. For most components made of the above-mentioned materials, the chief guideline is the weight. Other accessories such as pallets are designed to be used several times. We may assume the energy they require is generally negligible compared to that needed for the package itself. For example, a typical wooden pallet is made of 18.8 kg softwood, 6.65 kg hardwood, 1.16 kg nails and iron strips. The production data are summarized in Table 4.8. In total, only 308.10^6 Joule are necessary; once again, the transport of raw materials seems to be of minor importance (ca 2.5%) in the total amount. It becomes even more negligible if the pallet is used several times. Crown corks are made of tinned iron and the average weight per piece is 2.14 g. Crown corks are mar-

Table 4.8 Energy needs (in MJ) for the production of packaging accessories

Object and material	Production process	Electricity	Petroleum and derivatives	Other	Energy content
Pallet	18.80 kg (softwood)	–	125.00	–	438.00
(wood)	6.65 kg (hardwood)	–	44.00	–	–
	Sea transport	–	2.63	–	–
	Road transport	0.06	4.60	0.50	–
	Steel (nails)	2.50	3.82	20.00	–
	Assembling	105.00	–	–	–
	Total	107.56	180.05	20.50	438.00
Crown cork	Steel	1.31	8.65	46.97	–
1000 pieces	Tin	6.23	22.61	25.02	–
of 2.14 gram	Producing crown	4.03	8.25	0.15	–
each (tinplate)	Disposal	0.10	0.87	0.05	–
	Ink	–	0.03	–	0.06
	Coatings	0.26	4.23	2.61	5.69
	Adhesives	1.34	10.56	–	6.85
	Packaging	0.52	1.43	0.92	0.99
	Total	13.79	56.63	75.72	13.59
Closure	Aluminium	91.58	73.29	22.96	–
1000 pieces	Producing closure	6.40	15.97	0.32	–
of 1.31 gram	Disposal	0.05	0.29	5.59	–
each	Coatings	2.18	35.98	22.17	48.37
(aluminium)	Adhesives	2.29	17.97	8.01	11.65
	Packaging	0.75	2.05	1.33	1.43
	Total	103.25	145.55	60.38	61.45

[Source: Calculated from data, Boustead, Hancock (22)]

keted per lot of 6000 pieces in fibreboard returnable packages of 1 kg (reused three times). Manufacturing data are also mentioned in Table 4.8.

The energy required for the production of aluminium caps (soft drinks, mineral water, etc.) is, as expected, much higher despite the fact that their weight is half the weight of conventional crown corks. Aluminium caps are generally marketed per lot of 6000 pieces in fibreboard returnable packages of 1.2 kg (reused three times). It appears from these considerations that recycling those metallic components is particularly prudent as the energy needed for raw materials processing is saved every time. Also, the energy required must be calculated from the weight of the material used, because the forming energy is negligible. However, metal closures are increasingly being replaced by ones made of plastic.

4 TRANSPORT OF PACKAGING

At the beginning of this chapter, we emphasized the huge part taken by the transport of goods in total energy needs and on the various environmental nuisances involved. The popular press has stressed that the 'weight' of the packaging is the main factor influencing the transport cost, especially when interest groups are in conflict.

Every shopper has surely compared the effort to carry a crate (ca 1 kg) containing 12 litres water (12 kg) packaged in glass bottles ($12 \times 0.7 = 8.4$ kg) with the effort to carry the same quantity packaged in plastic bottles weighing 40 g each. In the former case we carry $1 + 12 + 8.4 = 21.4$ kg, in the latter only $1 + 12 + 0.5 = 13.5$ kg. This argument is relevant provided we carry our own purchases. In reality, supermarkets, accessible by car, provide trolleys to help customers take their purchases, even the largest ones, to the parking place without getting tired. Everyone has perhaps realized that the effort, the force necessary to pull burdens is but a fraction of the force needed to carry the weight in question. So the energy consumed in the transport sector does not depend as much on the weight of the

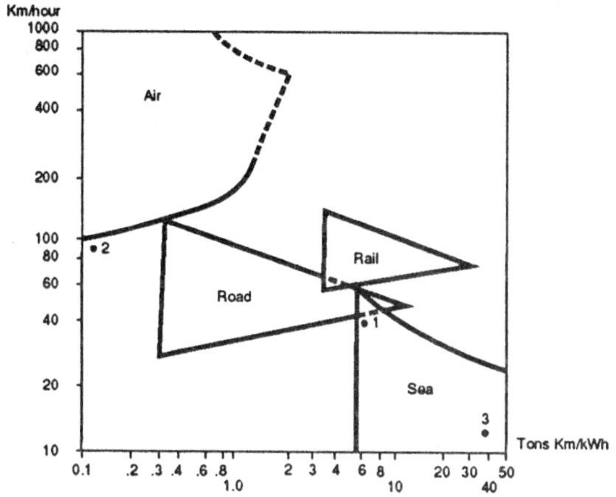

Figure 4.3 Energy efficiency of goods transportation as a function of the transport mode

carried goods, as on the speed, the distance and the method of transport. Figure 4.3 outlines the transport efficiency per unit of energy as a function of the speed (6).

It appears that sea transport is indeed slow but highly efficient. Rail transport is on average around three times faster (80 km/h). Depending on the size of the lorry, distances ranging from 3 to 15 km are covered per unit of energy (kWh) and of weight (1 ton). Equally, the faster air freight implies a much higher energy consumption.

The average energy consumption rate per ton and per km, expressed in kWh and in 10^6 Joule, is given for the different means of transport in Table 4.9. This shows that road transport becomes relatively cheaper by increasing the chargeable load. The following example illustrates how to use these data. To cover a distance of 1 km, a lorry completely loaded with 3 tons consumes an average energy equal to $3.45.10^6\text{J} \times 3 = 10.35$ MJ. For a distance of 100 km, $10.35.\ 10^6\text{J} \times 100 = 1035$ MJ are needed. This implies a consumption of approximately 22.85 litres of diesel oil (cf. Table 4.2). If the lorry covers this distance unloaded, it will consume on average 70% of the maximum, i.e. $1035.10^6 \times 0.7 = 725$ MJ (or ca 16 litres of diesel oil).

A lorry delivering 140 crates, each containing 12 litres of water packaged in glass bottles (0.7 kg per bottle), transports over a distance of 50 km a weight of 140 kg (crate) + 140 × 12 kg (water) + 1680 × 0.7 kg (packaging) = 2996 kg; the energy consumption amounts to 518 MJ. If the same number of crates containing empty bottles is taken back, the weight carried will be 140 kg + 1680 × 0.7 kg = 1316 kg and the consumption $518.\ 10^6 J \times 0.7 + (1316/3000 \times 518.MJ \times 0.3) = 431$ MJ.

This calculation proves that the additional energy consumed (68 MJ or ca 1.5 litres diesel oil) to carry back the empty returnable packages seems reasonable in comparison with the net energy saving of the glass packaging ($1.176 \times 15.69.10^9 =$ 18 450 MJ). Overall, we can say that the share of the packaging in the energy consumption for transporting of packaged goods is negligible compared with that needed to manufacture the packages. This statement is valid for all packaging materials provided the calculation pattern is applicable to the heaviest unit package.

Table 4.9 Average energy consumption rate and efficiency of transportation depending on the transport means

Transport system	MJ/ ton.km	kWh/ ton.km	Ton.km/ kWh	Ton.km/ MJ
Sea transport	0.087	0.024	41.10	11.50
Rail transport	0.406	0.113	8.87	2.46
Road transport				
1 ton	7.880	2.190	0.46	0.13
1-2 ton	4.590	1.270	0.78	0.22
3 ton	3.450	0.960	1.04	0.29
5-8 ton	1.650	0.460	2.18	0.60
10-12 ton	1.220	0.340	2.96	0.82
13-20 ton	0.770	0.220	4.65	1.29

[Source: Calculation based on Boustead (22)]

V

Packaging and waste

1 INTRODUCTION: PROBLEM OUTLINE

Life in general and our technical processes in particular involve the production of waste materials. Refuse from plants, animals and human beings, as discussed in Chapter II, can be 'eliminated' through a succession of biological processes.

As man has learned to process raw materials in order to produce new materials, and has expanded their production to ever larger scales, different kinds of wastes have been generated which can no longer be processed by the exclusive work of nature. Contamination and/or pollution problems generated by industrial processes have also been examined in Chapter II.

From one point of view, packaging wastes can be regarded as the solid refuse resulting from the use of goods. In Chapters III and IV, we paid attention to the material and energy aspects of the re-use and the recycling of packaging materials. Problems related to treatment and/or disposal of packaging wastes become more obvious if we split them into four categories: internal wastes, new wastes, old wastes and litter.

1. *Internal wastes*: unused packaging materials and/or wares rejected by the factory for breakage, damage or incorrect processing can be directly recycled. Glass refuse (cullet) in a bottle factory is a typical example: it can be directly sent back to the melting furnace. This is also feasible for paper in paper mills.
2. *New wastes*: these may include unused scrap, rejected packages or parts of packages, generated during their man-

ufacture, which cannot be recycled inside the factory but are collected from there directly. Steel plate scrap left in the manufacture of metal cans, unlike glass refuse, cannot be used in the can factory but can be pressed together and collected as fresh scrap to be sent back to metal furnaces.

3. *Old wastes*: this group includes all used packaging or 'post-consumer waste' remaining after usage of the packaged goods. This kind of waste differs from the first two groups by:
(a) its dispersion: the greater quantity does not come from one particular place but is widespread throughout the population. It is mainly rejected as household refuse in thewaste disposal bag.
(b) its heterogeneity: the fact that most types of packaging are not kept separate means that the household rubbish bin contains a 'complex' mixture of rubbish and packaging materials. For any recycling to be feasible (as argued when the plastic material is homogeneous), drastic sorting is needed. Selective collection of waste implies that every citizen must keep the different kinds of used packaging in separate bins. Glass, metal, paper and plastics are easily identifiable. Yet a sorting of metal, glass and plastics must be made: iron can be removed from mixed waste due to its magnetic properties, coloured glass can be separated from clear glass by its colour. In the case of plastics, sorting into different types is practically impossible because of their similarity, although one way is to provide plastic packaging with a colour code (cf. identification function page – or a material-specific embossment). Selective collection of glass packaging is practised. This is not valid for paper packaging: here newspapers and periodicals are principally involved. One objection against separate collection of used packaging is the space required for storage and the potential odour nuisance (cf. case study in Vienna, 177, 179). Collection (common or selective), sorting and cleaning cause the recovery of materials from household refuse to be an expensive operation. As already pointed out, raw materials prices are the major parameter in determining to which degree separation and recycling should extend.

(c) hygiene: used packages may not only be contaminated by traces of the packaged content but also be spoiled during transport, usage, or by the refuse mountain into which they are dumped. Depending on the household refuse water content (higher in winter than in summer time), micro-organisms will grow and microbiological degradations will set in where possible, collected paper and board will disaggregate, hence sorting will not only be more difficult, but the fibrous material (cellulose) will also lose its quality.

4. *Litter*: by this we mean used packaging people throw away carelessly. It spoils recreation grounds, public roads and parking areas (e.g. along motorways). In this case too, the variety of materials is wide (158). The estimated composition is 6% plastic, 6% glass, 16% metal and 60% paper. Because of the dispersion, the financial cost for the collection of litter is difficult to quantify: according to A.J. van der Kuil (168), as early as 1971 it was estimated at ca 10 000 BeF/ton. In some cases (beaches, roads), depending on the density, the wind can create local but unsightly heaps. Here paper will not be a problem for natural waste disposal as it is more easily degraded by water and biologically decomposed. For other materials (glass, metals and plastics) which are more resistant against these natural processes, the efficiency of the technological waste treatment methods is generally limited. This kind of 'litter' can be slightly reduced by providing appropriate rubbish bins and educating people to be more environmentally minded.

2 HOUSEHOLD REFUSE

According to OVAM, household refuse includes all solid wastes generated by normal households. Refuse from restaurants, shops and small factories can be added to it. In large towns, the solid wastes resulting from administrative activities (mainly paper) and coming from markets (mostly vegetable and fruit refuse) are picked up by the house-to-house collection.

Household refuse can also be split into two groups: on one hand, the materials put in the household rubbish bag or bin – called 'domestic' refuse – on the other hand, all materials which are not put in the rubbish bin because of their size but which are collected separately. The latter group is called 'bulky waste'. Respective shares of domestic refuse and bulky waste in the household refuse average 85% and 15%. Their quantity and composition are variable: they depend on the season, economic development of the region, technical development of the consumer goods sector, standard of living and finally on the social structure. The evolution of the household refuse quantity per capita and per year is given in Table 5.1; it evolves almost in parallel with the packaging consumption rate. Since the early 1980s, more and more attention has been paid to the separate collection of glass and paper: in 1988 in Flanders these averaged about 3% and 2.7% respectively of the total amount of refuse produced (see Table 5.2).

In order to achieve lower waste disposal rates in the future, investigations were also set up to evaluate the amount of plastic refuse available in the short- and long-term for further processing.

Table 5.1 Evolution of household refuse in Belgium during the period 1958–1989 (in kg per capita and per year)

Year	Household refuse quantity	Year	Household refuse quantity
1958	280	1972	360
1960	270	1974	370
1962	275	1978	340
1964	290	1981	286*
1966	340	1984	285*
1968	340	1987	306*
1970	345	1989	273**

*production rates in Flanders; ** data published by APME
[Source: OVAM]

The latest data for Belgium tend to indicate a decrease: in 1981 the household refuse production per capita and per year averaged 286 kg; this may be the result of the selective collections dating back to 1975. In 1981 in Flanders selective collections for glass and paper were already organized in 235 out of 313 municipalities; the quantity picked up averaged 7.9 kg per capita and per year. It is obvious that this amount is too small to explain the steep drop of the household refuse production; the economic recession seems to be a more probable explanation. The evolution of selective collections in Flanders during the period 1981–1988 is summarized in Table 5.2

Table 5.2 Evolution of selective collections in Flanders during the period 1981-1988

Year	Population	Production of household waste (in 1000 ton)	Selective collection of glass (in ton)	kg/capita glass collected	Selective collection of paper (in ton)	kg/capita paper collected
1981	5 629 004	1 614	–	–	–	–
1984	5 662 336	1 616	59 500	10.5		
1987	5 685 601	1 741	–	–	–	–
1988	–	2 300	53 200	9.4	42 500	7.5

[Source: OVAM]

The data of A. Buekens (24), N. Willems (183) and P. De Vos (in *Info-Leefmilieu* Jan. Feb. 1983) show that in the last five years the nominal quantity of packaging waste could have decreased but its relative share in the total quantity has remained almost constant (ca 28%). According to the data issued by J.A. Van der Kuil (168) and by the 'Stichting Verpakking & Milieu' (*Missets Pakblad*, 5 Aug. 1983, p 8), in the Netherlands during the period 1971–1981, the share of packaging waste steadily decreased from 32% (1971) to 25.2% (1978) and 22.4% (1981), compared with a small increase of the total household refuse quantity. The changes are as follows (quantities in thousand ton):

		1978	1981	1985*
Total household refuse		3610	3800	1681
Total packaging waste		910 (25.2%)	851 (2.4%)	348 (20.7%)
Total packaging of drinks		287.5 (7.9%)	291.5 (7.7%)	99 (5.9%)
including	glass	238.9	224.1	104.4
	cardboard	22.7	37.7	122.9
	plastic	8.8	9.4	76.40
	cans	9.2	12.7	44.50
	crown corks	7.0	7.3	–
	caps	0.8	0.29	–

* (Flanders)

To demonstrate once more how variable the composition of household refuse is, Table 5.3 reflects its evolution during the period 1932–1988. The share of packaging, based on these figures, is to be considered as a 'maximum': the heading 'paper' also includes writing paper, the heading 'glass' includes fragments of window glass, etc.

The drop in the ashes content and the vegetable waste is noteworthy; the former results from an intensive use of electricity and liquid fuels instead of solid ones, the decrease of vegetable waste is the consequence of the increasing usage of ready-cooked and pre-packaged products. The expansion of the paper and plastic shares contribute significantly to lower the density (1500 → 500 kg/m^3) and to raise the heat content of household refuse (800 → 2500 kcal/kg).

Rigid forms of packaging, such as cans and some plastics (PVC, PS), will always lead to a bigger volume of household refuse and, to a lower degree, to a decrease of its weight. Conversely, the heat content of paper (366 kcal/kg (dry)), PVC (4100 kcal/kg), polystyrene (9500 kcal/kg) and polyethylene (10 400 kcal/kg) will add to its calorific power.

Long-term forecasting about the properties of household refuse is uncertain, as its components will depend on their

Table 5.3 Evolution of the average composition of household refuse during the period 1932–1988 in Belgium (as percentage of the total dry weight)

Material	Year										Special properties
	1932 (1)	1946 (1)	1950 (1)	1965 (1)	1974 (2)	1977 (3)	1980 (4)	1981 (5)	1985 (6)	1988 (6)	
Paper and cardboard	21.7	6.3	11.9	29.6	35.0	30.0	22.0	22.3	28.3	20.9	(B) + (R)
Glass	2.3	2.6	4.2	2.9	7.0	10.0	11.0	8.1	7.5	7.6	R
Metals	2.7	4.0	2.2	4.2	5.0	4.5	4.0	3.0	3.7	3.7	R
Wood	4.0	3.7	2.4	4.1	4.0	–*	–*	0.4	0.6	–**	(B)
Ashes	30.2	31.7	36.4	21.3	23.0	6.0	–**	1.2	1.8	–**	–
Vegetables	34.0	42.1	32.3	24.0	15.0	43.0***	45.0***	51.7***	46.6***	6.8	B
Plastics	–	–	–	1.0	4.0	5.0	5.5	6.8	7.6	5.6	(R)
Various (e.g. textiles, rubber)	5.1	9.6	10.7	8.9	7.0	7.5	12.5	4.3	3.9	55.4***	
Packaging share (max.)	26.7	12.9	18.3	35.7	51.0	54.0	47.0	41.2	39.5	32.2	
Total R+(R)	26.7	12.9	18.3	34.7	47.0	49.5	42.5	40.2	47.1	37.8	
Total B+(B)	59.7	48.4	44.2	53.6	50.0	73.0	67.0	74.0	75.5	83.1	

* included in vegetables; ** included in various; *** including garden waste; B: biologically degradable; R: recyclable; (B): limited biodegradability; (R): limited potential recycling

[Sources: (1) de Neveux and Mesnil's report, *Techniques et sciences municipales*, 61, 1 (1966); (2) Personal data (106); (3) Willems (183); (4) Data for Flanders: OVAM's study (1982); (5) According to IVA's report 53: 10 years result; (6) OVAM] value as scraps. During World War II for instance, the share of paper in rubbish was small, perhaps because it was burnt by the consumer. Selective collections and strong trends towards recycling are other reasons why the composition of household refuse can no longer be considered as being constant. So the share of glass in household refuse dropped from 13% to 7% during the period 1974–1982 in the Netherlands; during the same period, selective collections of paper reduced its share from 26.1% to 21%. We cannot foresee which materials will disappear in the near future or which new ones will come onto the market and thus also contribute to the waste stream. From Table 5.3 we can assume that the heat content is subject to important variations. Nowadays, domestic refuse is in this respect quite comparable to lignite, a combustible used on a huge scale e.g. in Germany.

Besides the recyclable and/or combustible materials (packaging and others), domestic refuse also contains biologically degradable components, such as vegetable waste and food remains, as well as biologically stable wastes such as stones and ashes.

Combination with plastic reduces the biodegradability of packaging paper; the share of vegetable waste from food preparation follows a downward trend. Hence the biodegradability of domestic refuse is now lower than before. However, gardening has become more important, and this waste is also collected and able to biodegrade.

The nature of the materials used for modern household appliances which are also present in the bulky waste, makes its treatment difficult. This becomes obvious when studying its composition (see the case study illustrated in Table 5.4).

Nowadays household refuse may rightly be considered as 'urban ore', the 'exploitation' and/or 'processing' of which is not easy because of the potential and unpredictable fluctuations of its composition. The steadily increasing refuse volume on one hand, and the greater saturation of the available dumping grounds on the other, stress the urgency for a solution.

Table 5.4 Average composition of bulky waste (the Netherlands 1977)

Nature	%	Composition and subdivisions	
Paper and board	27.0	11.0	Packages
		16.0	Newspapers, periodicals
Metal	16.7	2.0	Bicycles
		3.0	Washers, driers
		1.0	Refrigerators
		2.0	Springs
Metal+plastic+glass	1.3	1.3	Radio and TV sets
Plastic+rubber	2.0	0.8	Buckets
Wood	23.0	2.0	Tables, chairs, etc.
		5.0	Garden refuse
Textiles	26.0	25.7	Floor covering
		0.3	Cushions
Miscellaneous	4.0	2.4	Soil

[Source: OVAM (1983)]

3 HOUSEHOLD REFUSE: PROCESSING AND ENVIRONMENTAL IMPACT

In the past, household refuse removal was of minor importance: it could be dumped on ground where no 'neighbourhood nuisance' would arise. Licensing would preside over any problems. With the principle of the closed natural cycle in mind, an ideal waste treatment method should not itself produce any 'waste'. Every important industrial material should be a wise contribution to:

- savings in natural raw materials,
- drop of the energy consumption rate,
- diminution of the environmental pollution induced by the exploitation, processing and transport of raw materials.

Finally, biodegradable products should return into the natural cycle through the agricultural sector. The existing processing methods are not at all ideal because:

- material and/or energy recovery is limited,
- the method and/or the treatment facilities are not free of pollution or nuisance,
- refuse (solid waste products) remain which must still be dumped.

These considerations lead to the thought that a wise combination of these methods should offer the best guarantees for environmental protection. The main treatment methods used in some West European countries are listed in Table 5.5.

It is indisputable that household refuse disposal has been improved lately due to the building of treatment facilities near the great urban centres and to the efforts made in this sector. Since the publication of the decree dated 2 July 1981 (Belgian State Journal, 25 July 1981) relating to waste management, and since the establishment of the 'Openbare Afvalstoffen Maatschappij voor het Vlaamse Gewest' – abbreviation OVAM (Public Organization for Waste Management of the Flemish Region) – by the RD dated 30 July 1981 (Belgian State Journal, 20 August 1981) major progress has begun in environmental politics.

Table 5.5 Comparison between household refuse treatment methods as a function of their importance (in % of the total quantity) in various countries

Method			Country			Average cost
	FRG (1978) 1989	GB (1978) 1989	DK (1978) 1989	NL (1978) 1989	Belgium (1972) (1978) 1989	per ton (in BEF) (1983)**
Landfill	67(76)	89(86)	28(45)	60(55)	(78) (58) 39	300
Incineration	29(21)	10(9.5)	70(25)	36(27)	(10) (25) 52	1070
Composting	4(4)	1(0.5)	2(35)	4(18)	(1.5) (3.3) 9	1000
Others	0(3.7)	0(–)	0(–)	0(–)	(–) (–) 0	

* Costs in 1988 were raised by 10% with respect to 1983; since 1987 an environmental tax of 3000 BEF/ton is in use
[Source: Belgian Packaging Institute]

OVAM has been operational since the publication of the RD dated 14 September 1981 (Belgian State Journal, 26 September 1981); its main function is to supervise the application of the decree's provisions and to stimulate waste prevention and material recycling. We must point out that the decree (art. 28) gave the municipal authorities the freedom to issue specific regulations; however, every treatment system must get an exploitation licence and this is applicable to any industrial activity. The permanent membership is empowered to grant licences and to fix the relevant conditions. According to article 5 of the decree, the 'abandonment of wastes' or, in other words, 'unauthorized dumping' is now forbidden.

During the last two years, great progress has been made by local authorities as well as on a European scale, to reduce the amount of packaging waste. The European directives coming into force in the year 2000 prescribe the following targets:

- 90% removal by weight of packaging materials from the waste stream for recovery,
- 60% by weight of each packaging material should be recycled,
- final disposal of packages should not be more than 10% by weight.

Meanwhile, the EC member states must work out intermediate targets and may adopt economic instruments to:

- promote the prevention of packaging waste,
- stimulate and encourage return of packages for re-use or recovery of materials and/or energy.

In several countries, a real cooperation between governmental agencies, industries and the users representing their federations and/or associations has come into force and agreements have been signed, mostly on a voluntary basis. For example, in the Netherlands the 'Convenant Verpakkingen' was signed 16 May 1991; in Belgium an association called 'PRO' (Prevention and Recycling Organization) was established 19 June 1991 to work out the objectives from the Convention of Packaging Wastes which was signed between the Flemish Government and industry on 26 March 1991. Similar conventions for the French and Brussels regions were also signed.

The German federal cabinet passed on 14 November 1990 a device 'Verpack VO' aimed at the prevention of packaging waste. This decree is better known by the name of the Minister for Environmental Affairs, 'Topfer'. In principle, this decree states an obligation to return packages and to charge deposits in the near future. The 'Duales Systeme' was established to make separate collection and waste treatment as efficient as possible. Also, private initiatives came into force by labelling environmentally friendly materials with a so-called 'Grüne Punkt' (green dot).

Up to now most attention has been paid to packaging materials, because they contribute more than 30% in volume to the waste stream and cause real visual nuisance when left carelessly in the environment. Expansion of these special packaging waste decrees may be expected in the future for other consumer products as well. Recovery and recycling programmes for plastics other than packaging will come into operation: they are recovered from buildings, old cars, from the distribution industry and from the agricultural and horticultural sectors. Most of these plastics become available as waste with a delay averaging 10 years: at the same time the consumption of plastics in different sectors is tending to rise continuously. During the period 1980–1990, for instance, the amount of plastics used in a car rose from 61 kg to 107 kg.

The plastic waste produced per sector for several countries is summarized in Table 5.6. It is clear that in 1989 about 20% of the total amount of plastic waste produced has been re-used, and this is likely to be extended in the future. Unfortunately this amount recovered as material and/or energy is only a fraction of the consumption.

Table 5.6 Summary of plastic wastes (in 1000 tons) produced per country and per sector in 1989 (percentage of plastic consumption)

Country	Total amount of plastic waste	Agriculture/ horticulture	Car industry	Building	Distribution industry	Household waste
Belgium and Luxemburg	316 (22.2)	23	20(10.0)	34	34	205
Denmark	117(24.9)	10	9(39.0)	13	17	68
France	1800(49.8)	72	115(31.8)	58	195	1360
Germany	2636(43.9)	162	132(23.3)	202	385	1755
Netherlands	529(47.2)	22	31(91.1)	35	51	390
UK	1830(53.1)	28	93 (44.9)	110	199	1400
Total EEC	10 332(45.0)	475 (40.0)	563(30.7)	544(10.6)	1220(70.0)	7530(57.5)
Total WE	11 433(44.1)	532(40.0)	620(30.5)	616(10.5)	1331(70.0)	8334(59.4)

Re-application of materials (reprocessing by thermoforming) in EEC 784 (1000 tons): 7.6%
in WE 846 (1000 tons): 7.4%

Recovery of heat content in EEC 1378 (1000 tons): 12.8%
in WE 1675 (1000 tons): 14.7%

[Source: APME, *Kunststof en Rubber* (1991)]

A specific publication (25) deals in detail with the various treatment methods. Hence we will only outline them here, paying attention to some aspects in connection with packaging.

a) Controlled and non-controlled dumping

By 'non-controlled' dumping, we mean waste disposal without any sorting or treatment. This method is unhygienic (vermin proliferation, pathogens dispersion, etc.). The presence of flammable products (packaging materials) combined with the potential 'magnifying-glass effect' of glass refuse creates a fire hazard during dry seasons. Packaging refuse, especially the lightest fractions made of paper and plastics, can blow away and be dispersed by the wind. Packaging, due to its characteristics, will contribute only to a very limited degree to the pollution of surface water by washing (174). Since 15 May 1983, non-controlled dumping has been forbidden in Flanders.

In the case of 'controlled dumping' (landfill), the refuse is regularly flattened with bulldozers and covered with a layer of earth. This reduces the most detrimental impacts. Licensing is required for controlled dumping. Both methods may be classified as among the cheapest; however, they do not lend themselves to material recuperation.

b) Incineration

Generally incineration plants burn household refuse without previous material recuperation at a temperature of ca 1000°C, sometimes with the addition of waste oil, biological (bio-industrial) or industrial wastes. The nature of the wastes and their largely diversified composition requires mobile grates specially designed to avoid any obstruction due to the sintering of glass, metals and stone scraps. As already stated, it is mainly the combustion heat of packaging materials (paper and plastics) which is recuperated in this way. As shown in Chapter IV, the conversion yield of heat into electric power is very low. Where household refuse is incinerated, part of the heat is used to dry off water and thus reduce that which is recoverable. Additionally in the conversion of the remaining heat (via steam) only 25% efficiency can be achieved.

Calculations (106) demonstrate that the incineration of all Belgian household refuse would cover only some 1% of the needs in electric power.

Of course, incineration generates sterile ashes (volume is highly reduced) which must finally be dumped. The average composition is as follows:

Silica dioxide (SiO_2)	30–50%
Aluminium oxide (Al_2O_3)	20–30%
Ferric oxide (Fe_2O_3)	10–20%
Calcium oxide (CaO)	7–15%
Magnesium oxide (MgO)	1.5–2.5%
Others	6%

The environmental problem is not so much the dumping of these wastes as the noise and the fumes and gases coming from the incinerator. Owing to their composition (see Table 4.3), combustion gases should be treated and/or purified before they escape through the chimney. Organic material will be converted into carbon dioxide (CO_2) and water. On the other hand, the presence of chlorinated (e.g. plastic resins, PVC, PVDC, Saran) and sulphur-containing (e.g. proteins) compounds will lead to the formation of corrosive gases; glass, metal and stone scraps will produce solid components (slags). The incomplete combustion and/or the warming-up of most plastics might generate some very toxic decomposition products. The necessary treatments (gas absorption by water, dust collection in electro-filters) produce effluents (which must be treated as well) and solid wastes to be dumped with the ashes. This brief survey explains why 'incineration' is very expensive (see Table 5.5). Nevertheless incineration is sensible when very specific flammable products are concerned, for which the most appropriate facilities can be designed.

c) Composting

Composting is a method to treat, in a controlled and generally quick way, biodegradable products by micro-organisms (bacteria, moulds, etc.). Heat generated during the 'biological' combustion (aerobic conditions) can provoke a rise in temperature able to kill the pathogenic bacteria. The respiratory quotient RQ is used as a check parameter: it gives the ratio between the produced quantity of CO_2 gas (in litre or mol) and the consumed quantity of oxygen gas (in litre or mol) (see also page 41). The stabilized result, called compost, is one of the oldest known products which can be used as fertilizer or soil conditioner because of its composition. The ratio carbon to nitrogen (C/N) in the final product is important: ideally it should amount to 30–35. Should the ratio be too low, paper refuse, straw or leaves should be added; should it be too high, the adequate ratio could be reached by removing paper refuse or by the addition of sewage sludge and fish or meat remains.

The C/N ratio for humus amounts to approximately 10–20. Due to the nature of the process, many packaging materials remain unaffected. Some (mainly iron and cans) are separated before composting. Generally glass, plastics, tin, aluminium and paper–plastic combinations (laminates) are mechanically removed after composting. Selective glass collections have significantly improved the compost quality. In Ghent, compost has already been prepared for 18 years according to the so-called DANO process. Household refuse is supplied at one end of a horizontal rotating cylinder (length 26 m, diameter 3.5 m); after ca 3 days, the compost mixture is discharged at the other end to be refined. The compost yield averages 53% (40). This waste disposal method furnishes some recyclable materials, the selling of which contributes to a reduction in the working costs. This share is small compared to the total quantity processed and the cost (174); the quality of compost is improving, however because of the separate collection of glass and some dangerous compounds. Removal and purification of packaging paper in order to recycle it is perhaps technically feasible (by batch methods), nevertheless it is non-profitable when compared to the recuperation opportunities offered by selective collection (non-packaging papers) and by the fresh internal production of paper refuse. So paper and plastic packaging wastes are incinerated and the heat is used for heating purposes.

For large towns, combined composting, selective collection and incineration should perhaps be considered as the most flexible disposal method. At one time, selective collections were a source of income for 'rag pickers' (24). This activity has been almost completely abandoned because the value of secondhand materials has remained rather stable in contrast to increasing living costs. It has not completely disappeared but it is more targeted than before on particular industrial refuse, the composition of which is homogeneous; as far as household refuse is concerned, it is even achieved on an industrial scale. Thus, we could think that selective collection (at home or in containers), like a special kind of sorting method, would disturb the balance of costs–benefits. Nevertheless, a case study on the sector Watermael-Boitsfort during the period 1975–1978, has shown that in this respect, selective collection of glass and paper is not showing a deficit.

Selective collection, various sorting methods and composting can hardly be considered as environment spoiling activities. Of course, the use of compost as a fertilizer for vegetable growing may be dangerous: the compost quality always depends upon the absence of small quantities of heavy metals (e.g. lead, tin) or on insecticide traces from sprays. Although it is of lesser importance for the gardening and agricultural sectors, nowadays compost still seems valuable for the establishment of parks and for horticulture.

d) Miscellaneous methods

Listed below are some disposal methods which are either still experimental, or which are applicable to specific cases.

'Pyrolysis' consists of heating organic products in the absence of oxygen: instead of combustion products, a range of decomposition products is obtained including combustible gases and condensable vapours that contain some interesting chemicals like those produced by dry distillation of coal.

During the pyrolysis of household refuse, a great deal of water is distilled; but as a source of chemicals it is not very profitable; whereas when homogeneous organic wastes, such as old rubber tyres and plastic packaging refuse, are treated, pyrolysis will generate several monomers able to be re-used after purification for the production of polymers or processing of chemicals. As the share of plastic packaging in household refuse is heterogeneous and unclean, this application is limited. We may also wonder whether this method is advantageous energy-wise: the production of plastics from crude oil always consumes approximately 50% of the raw materials in the form of energy. Pyrolysis also requires energy and the 'raw material' is partly lost (carbonization) so that finally the energy or material recuperation can become questionable (see also Chapter IV).

There is another method called the 'Tezuka' process where neither material nor energy is recovered. In this 'global method', the household refuse is pressed into bundles which are stabilized with cement or tar. The confinement of air considerably reduces any potential composting. These blocks can be used to pack marshes, to bed roads, etc.

4 CONCLUSION

These considerations show that the recuperation of packaging materials or energy from household refuse can offer only a marginal contribution to the solution of raw materials and energy problems.

In large towns, waste disposal treatment has become a necessity. The heterogeneous and variable composition, the diversity and spoilage make the sorting of packaging materials into types as well as their recycling difficult. As far as packaging materials present in household refuse are concerned, only glass and metals have an importance for materials recycling. Packaging paper and plastics are more efficiently valorized in the form of combustion heat (energy recovery).

Selective collection or gathering of glass in containers and recycling is more expensive in terms of energy than the re-use of containers, nevertheless it is sensible nowadays because of the high labour cost involved by the organization needed for returnable packaging systems. Therefore a combined household refuse disposal method seems the most practicable solution, featuring a minimum risk of environmental pollution and of raw materials squandering. First the biodegradable components should be separated from the non-biodegradable ones by composting. If recyclable glass and metal are then removed from the non-composted material (plastic etc.), the remainder is suitable for heat recovery by incineration.

VI
Safety and toxicity of packaging

In this chapter we discuss some of the aspects of packaging which concern us all: as almost every product is packaged; everyday contact, handling and use are unavoidable. In many respects, there is always some slight uneasiness about the use of new materials. Nevertheless, we must trust the technology and the relevant legislation, especially about the packaging of foodstuffs. In the literature, all the hazards involving the use of packaging are grouped under one heading like 'potentially dangerous' and are studied by comparison with other hazardous events, working or living conditions. As far as packaging is concerned, the elements constituting a potential 'danger' can be classified on the basis of the following questions:

- Are the various types of packaging dangerous as everyday articles or, in other words, how safe is their use or can they cause accidents?
- Are the various forms of packaging in the food and pharmaceutical sector sufficiently sterile or, how safe are they against the penetration or contamination by pathogenic bacteria; does the packaging itself carry harmful germs?
- Could the various packaging materials transfer toxic or poisonous components into packaged foodstuffs and pharmaceuticals?
- Do the packages offer enough protection against accidental misuse of the content (e.g. should child resistant closures be provided)?

- Are the packages 'tamper-evident' (do they provide protection against deliberate or malicious contamination of content)?

1 PACKAGING IN EVERYDAY USE

Modern society teaches us how to live with risks. From childhood, we learn how to handle everyday articles in order to avoid accidents. Although technological progress has made buildings, vehicles and household appliances safer, many precautions must still be taken. Almost everyone faces a steadily reinforced dependency upon the goodwill of other people: we 'must' stop at a red light, an act we trust other people will also do!

Thus, each society takes some risks upon itself. Teaching the present throwaway society a behaviour code implies that people know how to use packaging properly. The main 'bad' property of glass is its fragility and it is unnecessary to demonstrate that glass fragments can cause severe wounds. Carelessly abandoning and/or smashing of glass bottles on beaches or on recreation grounds can be considered as a kind of aggression: only education can alleviate this. Used metal cans too have thin sharp edges apt to cause injury. In this respect, plastic packages are safer than glass or metal ones: generally they are flexible and do not break, hence sharp and/or dangerous edges are infrequent. Table 6.1 lists the number of accidents in the EC countries (1974–1978) due to the use of household articles and appliances, to food poisoning and to traffic, which caused either severe injuries or even death. This table shows clearly that packaging rarely causes injury and is never the cause of death.

Table 6.1 Average death and injury casualties per year, recorded in the EC countries (1974–1978)

Cause	Number of deaths	Number of injured and/or sick people *
Road accidents	52 442	1 552 000
Food poisoning	160	44 000
Food packaging (glass and metals)	nil	2–3000

* Only accidents which cause more than 24 hours stay in hospital are taken into account
[Source: Katan (90)]

2 PACKAGING FOR THE PROTECTION OF FOOD, ETC.

Among the functions packaging must provide is protection against insect attack and microbial spoilage, especially with foodstuffs. Penetration of the package can occur either through small holes resulting from manufacturing defects, through slits or cracks caused by folding, bending or deformation of the packaging material caused by incorrect sealing such as gaps in the glue layer in the case of paper bags, or imperfections during heat-sealing of plastic and the joints in metal packaging. Cavities between two badly joined packaging parts or the fluting in corrugated fibreboard, for instance, may become a repository for insect eggs or microbes. Such defects in the bulk packages for products such as fruits and vegetables can lead to spoilage of the packaged goods through rodent and insect attack. The risk depends on the nature of the foods themselves as well as the kind of packaging. For example, the more permeable to food aromas the packaging, the greater the probability that rodents and insects will be attracted. Hard materials with smooth surfaces lower the probability that they can find a weak point for attack. Hence polypropylene (OPP thickness 30 μm) and polyester (PETP thickness 50 μm) offer a better resistance to rodents than polyethylene (LDPE), cellulose film or polystyrene (PS thickness 50 μm). Polyvinylchloride (PVC thickness 60 μm) without plasticizers seems to be resistant against boring insects such as Rhizopertha domenica. By their nature, metal and glass packages are 100% safe against such attack, and by coating paper with plastic, for instance, it is also possible to increase the resistance of the paper. Infestation can generally be avoided by the use of insecticides, by regular inspection of the sterilization equipment and by the cleanliness of the premises and the staff.

Pyretene mixed with 10% piperonylbutoxide can be used as insecticide. To limit the permeation of the sprays into the foodstuff in the case of long storage, appropriate combinations can be elaborated, such as those used for dried soups and for cocoa powder: a typical laminate might be parchment (40 g/m^2), adhesive (11 g/m^2), aluminium (thickness 9 μm), ionomer (23 g/m^2); after 12 months no permeation has been observed. Such precautions are especially important for hot climates with high humidity, where the contamination risk is higher. Variations in temperature to which packages are exposed during transport and during their shelf-life can also cause condensation of water vapour on the internal surfaces, creating a favourable milieu for the growth of micro-organisms, resulting in faster deterioration of the goods. Food hygiene must necessarily be in harmony with hygiene of the packaging materials; the re-use of bags, baskets, boxes, etc. requires precautions in warehouses and storage to avoid infestation. Removal with methyl bromide (CH_3Br) or phosphine (PH_3) is one method used to control insects. Tablets giving off 'phosphine' are placed in bags of groundnuts (peanuts) and lemons are treated with diphenyl. The lethal dose for insects must always be present without reaching a toxic level for consumers. The use of carbon dioxide and nitrogen mixtures to provide an oxygen-free atmosphere (MAP), a brief heating period to over 58°C or storage at low temperatures (0–5°C) can significantly slow down microbial attack. Ionizing radiations (200–500 Gy) also kill microbes and have sterilizing properties. Generally the contamination of foodstuffs after they have been packaged results from poor handling in transport and display.

Other harmful effects do not result from external factors acting on the packaged product but from contamination of the packaging materials which occurred before the goods were packaged. During manufacture, storage and transport of empty packages and during the actual packaging process, infection by bacteria and moulds can occur. As they are manufactured at a high temperature, glass packages are absolutely sterile as they leave the furnace. Contamination can take place only during cooling and/or storage. Plastics are shaped and made at lower temperatures (100–150°C), but the sterilizing effect of this brief heat-processing seems sufficient. Infection by moulds and bacterial spores is by air borne micro-organisms; loaded with dust particles and germs, air can contribute significantly to the infection of plastics particularly where electrostatic charges by friction are present. Storage of packaging near potential infection sites (e.g. meat slicers, bread, cheese) can increase their microbial content. Heiss (77) reported that inspection of 250 packages in 16 bakeries revealed on average 1 to 6 germs per dm^2 (see Table 6.2).

Table 6.2 Number of germs per dm^2 observed on bread packagings*

Packaging material	Average value	Maximum value
Paper treated with wax	6.2	50
Cellophane	1.4	7
Polyethylene (LDPE)	2.1	10
Polypropylene (PP)	2.0	7
Polyvinyl chloride (PVC)	1.7	–
PVC shrink wrap	4.9	21

* Area of packaging needed for an 800 g loaf of bread = 10 dm^2
[Source: Heiss (77)]

Lack of hygiene among the staff can contribute to the contamination by E. coli and staphylococci. To wrap non-sterile foodstuffs, 'clean' packaging materials are a must: they need not necessarily be sterile. Meat and meat products, such as patés, always have a surface contamination of the order of 104 to 105 germs per cm^2 (39). The germ content of clean polystyrene bowls however averages 'only' a few units per cm^2 which is negligible for the packaging of fresh meat products. Pasteurized milk contains on average 103 germs per litre; a few germs per dm^2 of packaging surface, E. coli excepted, is still acceptable here considering the short preservation period of a few days or a week.

Likewise, it is not reasonable to require germ-free packaging materials for vegetables and fruits. There are of course limits; for pre-packaged foodstuffs these are set at 10 up to a maximum of 50 germs per dm^2. Principally enteric bacteria are not allowed; in a word, the manufacturer must be able to guarantee the microbial quality requirements for foodtuffs fixed by the relevant legislation.

Conversely, in the case of 'sterile' products the requirements relative to foodstuffs and to their packaging must be strictly observed: 1 germ per dm^2 for a glass bottle contaminated during the cooling process is not acceptable for the packaging of long preservation (3 months) sterilized milk. The following example illustrates the requirements for cartons for UHT-milk packaging. Consider as acceptable a reject rate of one complaint per thousand 1 litre packages, then in every 1000 packages no more than 1 germ may appear on an internal surface, assuming the milk itself is free of germs. To avoid further risks and for safety reasons, a reduction factor at least equal to 10^5 must apply to empty 1 litre packages containing on average ten germs (1 to 2 per dm^2, i.e. 6 to 12 per 6 dm^2, the theoretical area of 1 litre volume being ca 6 dm^2). To obtain this germ-killing on high-speed packaging machines, a thin film of an aqueous solution of 30% hydrogen peroxide is sprayed on the plastic-coated paperboard sheets cut to shape. This film is then removed by blowing air at 80°C over each sheet. For the milk itself, a reduction factor at least equal to 10^8 is applied if we start with a pasteurized milk containing 10^4 germs per litre. The UHT process is used, therefore: this method consists in heating the milk rapidly (2–4 seconds) to a high temperature (130–150°C).

Thus, in the food sector, packaging has great importance as a shield:

1. against penetration by insects and vermin
2. against contamination by micro-organisms (post-infection of the products).

Hence the packaging surface itself must be free of germs especially in the case of 'sterile' goods. For the packaging of 'non-sterile' goods, plastic resins, the latest plastic packaging materials, require in this respect only the classical precautions. For the packaging of 'sterile' products, most types of plastic packaging, unlike glass or metal, cannot be sterilized after filling by a heat treatment. As mentioned earlier, post-sterilization by irradiation with ionizing radiations, is not always applicable because the necessary dose may modify the flavour of the products and/or change the character of the plastic packaging (modification of the migration behaviour after irradiation) (75, 113).

Hence, if necessary, glass and metal packages can be sterilized after filling and closing, which also tests the correct sealing of the packages. As far as plastics are concerned, pre-treatment sterilization will preferably be performed. We must consider here the various techniques to get an overview of any potential secondary effects.

Sterilization of empty plastic packagings, or of plastic materials in general, with hot air, steam, γ-rays are efficient methods, but the treatments could modify the colour as well as the tensile strength of the materials (176).

In the case of sterilization with dry hot air the best germ eradication is obtained as follows: 160 minutes at 160°C or 60 minutes at 170°C or 30 minutes at 180°C. These periods are long because the heat transfer is slow in the case of hot air. Few plastic resins, such as Nylon 6 or Teflon, will resist this high temperature, and these are not used in the packaging sector; most others distort and/or melt.

Sterilization with steam requires 20 minutes at 120°C. Here too, polystyrene and the useful polyesters will buckle; polypropylene will become milky white, etc. Such colour

Table 6.3 Modification of appearance and of tensile strength after irradiation of plastics with γ-rays

Polymer	Visual change after 1 sterilization at A, B, or C	Change in tensile strength (%) A	B	C
PS	Unchanged	91	118	59
PP	Light yellow/opaque	92	93	78
Plexiglass	Yellow brown	46	57	30
Nylon 6	Light yellow	40	89	56
Nylon 66	Yellow brown to beige	81	87	98
PETP	Light yellow to brown	99	93	95

A: dose 25 kGy at 1.6 kGy per hour; B: dose 25 kGy at 4.0 kGy per hour; C: dose 50 kGy at 4.0 kGy per hour

changes indicate structural modifications. For sterilization with γ-rays, 25 kGy dose is recommended; as shown in Table 6.3, colour changes will generally occur. Table 6.3 also indicates the change in tensile strength as a function of the irradiation dose. Note also that irradiation at lower levels induces greater modification of the tensile strength: this happens because the oxygen in the air has enough time during the irradiation to diffuse into the plastic and cause oxidation. With PVC, hydrochloric acid is rapidly produced, inducing some secondary reactions; moreover PVC tends to discolour very easily (113).

Sterilization with β-rays (electrons) is less damaging for plastic packaging but their penetration depth into the food is less. UV-light can also be used due to its germ-killing effect (by generation of ozone from atmospheric oxygen). In practice however, the available sources do not seem to have a sufficient germ-killing power to be used at the high packaging speeds which are taken for granted. Chemicals in the form of gases (e.g. ethylene oxide) or in solution (e.g. 30% hydrogen peroxide in water) are also suitable.

An important point in ethylene oxide sterilization is its toxicity and inflammability: special safety standards must be observed during the treatment; moreover all gas residues must be carefully removed. The quantity of ethylene oxide in the polymer after sterilization ranges from 100 ppm in polycarbonates to 50 000 ppm and even more in plasticized PVC. The gas can also cause the formation of toxic substances. A typical example is the potential reaction with hydrochloric acid (HCl) taking place in PVC due to thermal changes during forming or under the influence of light. The toxic 2-chloroethanol is formed as follows:

$$\underset{\diagdown O \diagup}{CH_2 - CH_2} + HCl \rightarrow HO - CH_2 - CH_2 - Cl$$

The oxidizing action of the hydrogen peroxide, due to its decomposition into water and oxygen, is less detrimental to the properties of plastics because the treatment is short and only affects the surface.

This leads to the conclusion that several sterilizations (in the case of possible re-use of materials) and/or utilization of a sequence of sterilization methods will involve, in the case of plastics, many unpredictable secondary effects, mostly damaging and possibly toxic.

Thus it seems reasonable to avoid any microbial contamination, by using, for example, clean germ-free packaging surfaces as well as by strictly controlling the hygiene of the personnel and equipment.

3 SAFETY OF PACKAGING IN RELATION TO MIGRATION

a) Definitions

In the literature, 'migration' refers to the potential exchanges between packaging and packaged contents. The meaning generally assigned to 'migration' is the transfer of matter from the package into packaged foodstuffs, apparently a one-way transfer. However, in most cases the phenomena is the result of material transfer in both directions. The assimilation commonly made between packaged product and foodstuff has been induced by the importance of the latter for public health; the migration phenomenon is general and applies also to packaging for pharmaceuticals, chemicals, detergents, cosmetics, etc.

Since plastics are used to package foodstuffs, to manufacture materials designed to be in contact with food, to make surgical prostheses, synthetic gut and tubes, as well as implants or items designed to be in contact with body tissues, many publications have drawn attention to the potential toxicity of these materials caused by interactions. Accidental contamination of goods caused by the opening and/or the use of packagings are sometimes ascribed to 'migration' phenomena (23); but packaging fragments such as fibres, bits of foil or glass fragments get into the packaged product by the action of mechanical forces. The occurrence of these accidents must of course be taken into account, but are not considered as migration in these discussions. The material transfers we study here result only from chemical and physical laws.

The components leaving the packaging surface are called migrants; a difference is made between global migration, being the sum of all migrants, (mostly unknown), and specific migration or the transfer of a specific component such as metal ions or plasticizers.

To protect public health, legislators have enforced quality criteria applicable to all articles in contact with food. The approval of packaging for food in Belgium started with the series of Royal Decrees dated 12.09.1972 and 25.08.1976, and was one of the most advanced and yet limiting legislations relating to the packaging sector. Harmonization at European level started with Directive 76/893. Limitations apply mainly to global migration, which may amount to a maximum of 10 mg/dm^2 contact area or 60 mg per kg of packaged product. This applies to all components, even those which are not toxic or have no physiological impact. The concept of setting a global migration limit is to avoid the necessity of determining the concentration of all migrants, whether they are known or can be identified.

Determination of specific migration is a more accurate,

time-consuming and expensive study; conversely, determination of global migration is cheaper, does not of course give specific information, but is still important because all packaging ingredients are enumerated in positive lists with the indication of the maximum allowable migration value. Determination of global migration serves as an indicator: if the test gives too high a result, possible toxicity may exist and when the maximum allowable global migration values are exceeded, specific migrations should be investigated.

Although migration has only been studied scientifically in detail for the last two decades, the phenomenon has been known for a long time; for instance, wooden barrels have been used to store and to give whisky its 'taste'; cork too gives a specific taste to wines and beer. It is also known that glass jars in which pickles have been preserved keep the smell of these packaged goods for a long time even after flushing with water: this indicates also potential absorption and/or exchange of components through the glass surface. Plastic boxes kept closed for a long period generally feature specific odours depending on the substances which have diffused through the surfaces.

b) Packaging classification

Depending on the way the migration occurs, packaging materials are classified into three groups, namely:

Class 1

Packaging which shows no migration phenomenon. This ideal situation can exist for various types of packaging provided the temperature and the nature of the packaged product have been taken into account. Ideal zero migration conditions can be met with solid (dry) products (flour, sugar, coffee, salt, dried vegetables, dried fruits, etc.) in contact with inert materials (glass, tinplate, paper) in a water-free atmosphere. Some plastics, such as nylon and to a certain degree polyethylene and polyesters, epoxy resins (for can coating), and cellulose film may be included in this class when they are used to package solid, dry and oil-free products. The ideal combinations relevant to the migration phenomenon are more theoretical than practical: e.g. fine sugar cannot be packaged in glass because of the fragility and weight of the packaging; as far as exchange possibilities are concerned, paper and board are equally suitable if the atmosphere is dry. Untreated and/or upgraded papers and boards are not adequate for packaging liquids or water-containing products. Likewise, unprotected metal cans cannot be used for acid liquids (vegetable or fruit juices) because of the specific reactivity of metals and metal oxides (the protective layer formed on tin and aluminium) with acids. In the case of liquid products (basic reactants excepted), glass seems to behave as the ideal packaging material inducing zero migration. For strongly acid solutions, some metal ions, such as calcium and sodium, can be exchanged for hydrogen ions from the acid; as far as acid foodstuffs (< pH 4.5) are concerned, this exchange is negligible, and moreover innocuous, as they are present in most foods (cf. Table 2.3).

Plastic films for frozen products can also be considered as 'ideal' packaging materials if the freezing takes place quickly and the packaging is removed rapidly after defrosting.

Class 2

Packaging inducing 'independent' migration, i.e. not caused by the packaged product. Most thermoplastics used for packaging belong to this class. Migrants are volatile components, especially monomers from the reaction medium used to produce the polymer, or generated by thermal breakdown during the forming of the polymer into a container (58, 60, 61, 62, 63).

The migration of these components takes place by a diffusion process. The volatile molecules will diffuse toward the surface of the plastic and there evaporate into the air. Because of the potential toxicity on the one hand (e.g. vinyl chloride, styrene), or of potential modifications in the organoleptic properties of the packaged foods on the other, those migrations must be limited: the RD dated 25.08.1976 and the EC directives 82/711 and 86/388 give maximum values for the migration of some monomers: the main ones are referred to in Table 6.4.

Table 6.4 Maximum allowable migration of some monomers into foodstuffs (RD 25.08.1976, Belgian Official Journal 24.09.1979 – RD 25.04.1980, Belgian Official Journal 18.07.1980)

Monomer	Relevant polymer	Max. migration in mg per kg product or mg per 6 dm^2 pack. area	Max. concentration in the plastic in weight (%)
Acrylonitrile (l)	PA	0.3	–
Ethyl acrylate (l)	PEA	2	–
Methyl acrylate (l)	PMA	2	–
Methyl methacrylate (l)	PMMA(1)	30	–
Ethylene (gas)	PE	60	0.1
Propylene (gas)	PP	60	0.1
Styrene (l)	PS	60	0.1
Terephthalic acid (l)	PETP(2)	30	–
Ethylene glycol (l)	PETP(2)	0.5	–
Vinyl acetate (l)	PVA	60	0.1
Vinyl chloride (gas)	PVC	0.01(10 ppb)	10^{-4}(1 ppm)
Vinylidene chloride (l)	PVdC	60	0.1

(1) plexiglass; (2) PET = polyethylene terephthalate; (l) liquid

The RD dated 25.04.1980 on migration simulation conditions has fixed strict requirements for vinyl chloride monomer while 1 ppm may be found in the plastic, no vinyl chloride may be detected in the foodstuff. This means that the permitted concentration in the foodstuff must be lower than the detection limit of the gas, namely 10 ppb (10 parts per billion).

A PVC bottle, used to package mineral water or oils, weighs on average 35 g; if we suppose that the monomer concentration reaches the maximum allowable (1 ppm), then the complete package may contain 35 μg vinyl chloride. Gilbert and Giacin (58, 61, 62, 63) succeeded in proving that all the monomer does not pass completely into the food: part of it remains in the polymer network due to Van Der Waalse's adsorption forces. Moreover, the quantity of diffused monomer is divided into two phases: the gaseous phase over the packaged content (head space) which escapes when the bottle is open, and the part in solution in the liquid medium. The difference in toxicity between vinyl chloride inhalation and ingestion by mouth is not clearly stated in the literature; if we suppose both effects are equal, the potential doses the consumers are exposed to can be estimated as follows: within three months, ca 20% of the total monomer quantity can pass into the packaged content, e.g. ca 7 μg: this quantity is thus below the allowed limit values. After a period of one year, it amounts to 40% or 14 μg (98, 171). Hence it seems in extreme cases (see also migration parameters) the migration limits might be reached. As it can happen that mineral waters and oils may stay in their packaging for more than three months, smaller packages could be used to reduce preservation (by the manufacturer or the consumer) during long periods. Nevertheless the capacity of most PVC bottles is more than 1litre (e.g. 1.5litre), the justification being that global migration is lower, and thus more favourable, in the case of big volumes than for small packages. The vinyl chloride quantity ingested each year by a consumer drinking daily one litre of water packaged in PVC bottles, amounts to a maximum of 10 μg × 365 = 3650 μg or 3.65 mg. This dose is approximately 10 000 times lower than the allowable annual dose 500 mg × 70 = 35 000 mg (body weight 70 kg).

The risk of death involved by ingesting vinyl chloride is considered negligible, namely approximately 3×10^{-13} (cf. Table 6.5). The same approach applies to other plastic monomers such as ethylene, styrene, etc. The quantity migrating spontaneously into food remains negligible in all cases (83, 124).

Spontaneous migration of liquid and/or solid components from packaging is even lower than that of volatile substances and can thus be neglected.

Class 3

Packaging where 'dependent' migrations occur, thus induced by the packaged content. These exchange phenomena happen with most packaging wares: the description of Class 1 materials

Table 6.5 Comparison of death risks

	Risk in one human lifetime	Risk per year
Occupational risks:		
Miners (silicosis)	$2.5\ 10^{-1}$	10^{-2}
(accidents)	$8.0\ 10^{-2}$	$3.0\ 10^{-3}$
Pilots (some accidents)	$6.0\ 10^{-3}$	$2.0\ 10^{-4}$
Firemen	$3.0\ 10^{-2}$	10^{-3}
Metal workers	$5.0\ 10^{-3}$	$6.0\ 10^{-4}$
Workers in PVC factory (exposed to 0.5 ppm monomer concentration in the air)	10^{-5}	$4.0\ 10^{-6}$
General risks:		
Smoking*	–	$5.0\ 10^{-3}$
Ionizing irradiation*	–	$3.0\ 10^{-6}$
Lightning*	–	$4.0\ 10^{-6}$
Meteorites*	–	$3.0\ 10^{-9}$
Vinyl chloride*	–	$3.0\ 10^{-13}$

* American statistical data
[Source: Van Vucht (171)]

has already shown that the occurrence of 'dependent' migrations is governed by the liquid contact medium and, to a lesser degree, by the temperature. The phenomenon depends on physical and chemical properties of both elements in contact with each other, on the kind of packaging and on the nature of the packaged product.

In basic or acid media, glass can exchange sodium and calcium ions: in the case of lead glass wares (such as crystal or some coloured glazings for ceramic wares), lead and/or cadmium ions can be exchanged with oxygen ions from the aqueous medium. In the case of glass wares, the allowable migration values for lead and cadmium amount respectively to 1 mg/dm^2 (5 ppm lead) and 0.1 mg/dm^2 (0.5 ppm cadmium). For ceramic packages and jars, the following migrations are allowed: lead 2.5 ± 0.25 mg/l (i.e. approximately $2.5/6 = 0.4$ mg/dm^2); cadmium 0.25 ± 0.025 mg/l (i.e. 0.04 mg/dm^2). In contact with water (pH 7) and oils, glass behaves like a neutral material (migration = 0). Tinplate (tinned iron) and aluminium in contact with acid or basic aqueous liquids will be corroded according to the specific chemical properties of the metal; the result of this corrosion will bring metal into solution in the form of metallic ions. Also, by bleaching (of vegetable juices for instance (122)) and the organoleptic modifications induced thereby, a foodstuff could become toxic depending on the kind of metal. The laws dated 04.04.1974 and 25.08.1976 enforce certain requirements relative to the purity and allowable migration for metal packaging: these values can be found in Table 6.6.

Table 6.6 Requirements relative to the purity of tinplate used to manufacture cans and maximum allowable values for the migration of metallic ions into foodstuffs

Tinplate components	Chemical symbol	Composition in % (1)	Max. migration (mg/kg food) (2)
Tin	Sn	99.7500(min)	1.00
Antimony	Sb	0.0400(max)	0.05
Arsenic	As	0.0500(max)	–
Bismuth	Bi	0.0150(max)	–
Cadmium	Cd	0.0001(max)	–
Copper	Cu	0.0700(max)	10.00
Iron	Fe	0.0150(max)	–
Lead	Pb	0.0500(max)	0.01
Nickel	Ni	0.0100(max)	–
Cobalt	Co	0.0100(max)	0.10
Sulphur	S	0.0100(max)	–
Zinc	Zn	0.0050(max)	–

(1) RD 4 April 1974, Belgian Official Journal 29 May 1974 pp 7723–7733; (2) RD 25 August 1976, Belgian Official Journal 24 September 1976 pp 12029–12082

As explained in Chapter III (see page (107), metals will pass faster into the solution as ions due to electron reactions when, in an aqueous medium, they come into contact with nobler metals. In the case of tinplate, if both metals (iron and tin) make contact in an aqueous medium because of scratches or cracks, iron ions will pass into solution due to the galvanic effect. This can result in perforations in the can. At the same time hydrogen gas appears on the tin layer, which could lead to explosion of the can under the influence of pressure. Due to the presence of metal ions, the flavour will change and the content could become toxic. In the case of galvanized iron (zinc-coated iron), the presence of zinc will prevent iron from rusting even if there are some cracks in the protective zinc layer.

Although these phenomena are well known, the relevant legislation is quite recent: the RD dated 15 March 1982 concerns damaged cans. This RD is an addendum to the RD dated 3 January 1975 concerning the noxiousness of materials and gives the following criteria for the rejection of metal food containers:

1. The can is rusted or is leaking.
2. The can is damaged by dents or other defects so that it could become permeable to gases, liquids or micro-organisms.
3. The flat surfaces of the can have been made convex.

Plastic materials are increasingly used to package foodstuffs and pharmaceuticals. The materials are new, the relevant mechanisms are but partly explained and studies are still in progress. Potential interactions between contents and packaging surfaces can be considered as a combination of adsorption phenomena linked with structural modifications of the surface which are followed by some diffusion phenomena: first, the plastic surface absorbs the liquid in contact and/or the dissolved components that can penetrate the plastic surface. Thus, the molecular distance between polymer molecules will increase and the various additives placed between them will become more easily 'accessible' and 'removable'. The contacting liquid penetrates into the surface to a depth ranging from only a few tenths to a maximum of a few microns. The volume of this area saturated with liquid increases, as can be observed under the microscope. Depending on the nature of the contact liquid (oily, aqueous, acid or alcoholic), limited or large additive extraction will occur. Like the monomer diffusion, the driving force here is the difference in concentration inside and outside the packaging surface. As several phenomena may take place simultaneously – penetration of contact liquid and extraction of additives – an equilibrium would theoretically be reached only after a long contact period. In practice it has been observed that the state corrresponding to the maximum possible global (or specific) migration, which varies of course with the temperature, is approached within a few per cent in five days (see Figure 6.1).

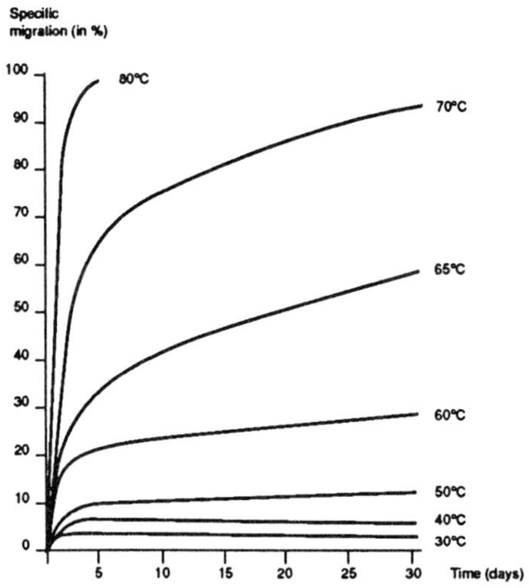

Figure 6.1 Migration as a function of time and temperature

Another important aspect is the need for an overview of the possible migrant components of plastic resins.

c) Additives for plastic resins

Generally 'plastic resins' may contain some ten different components. They can be classified as follows:

Reactant medium traces

Most plastics used in the packaging sector are produced on an industrial scale; the monomers react chemically (formation of high molecular compounds) in a liquid medium under the influence of catalysts, temperature and pressure. As solid matter is produced by this reaction at room temperature, components formed in the reaction can remain enclosed after separation and purification of the final product. Moreover solvent, catalyst traces, initiator and impurities in the reactants used, are included in this group. Where peroxide derivatives are used as catalysts, the migration expressed in mg active oxygen should not exceed 0.05 mg per 6 dm^2.

Intentional additives

This group includes substances incorporated in the polymers in order to obtain the properties required for the final packaging material or to facilitate its processing. They can be classified as follows:

1. *Property-improving substances*: among these, plasticizers are well known; they are added to give the material the desired flexibility. The choice of the plasticizer and of the necessary quantity depends very much on the kind of plastic, on the purpose (food sector, bundling of units, etc.) and on the desired softening glass transition temperature (see Chapter III, Plastics). Most plasticizers are liquids of high boiling point and low vapour pressure; they are inserted between the polymer molecules and thus reduce the crystallinity. Hence the friability and the fragility at the temperature of use (mostly room temperature 20°C but also freezing temperature –18°C and below) are decreased. The most common examples are esters of phthalic acid namely DBP (dibutylphthalate) and DEHP (di(2-ethyl n-hexyl)-phthalate)) added up to 20% by weight in PVC. An important compound representative of phosphoric acid esters is TPC (tricresylphosphate). Normally no plasticizers are incorporated in PE resins; hence, smaller molecules (with low molecular mass) act as plasticizer for the

bigger ones (with high molecular mass). This group includes also the usual colourants: they give the material a uniform colour. Solid and rubbery as well as fibrous products can also be used to influence the mechanical properties.

2. *Stabilizers*: these substances are added to protect the plastic against the action of light (especially UV-rays) and atmospheric oxygen. Several thermal treatments are performed throughout the different steps from the polymer preparation to the final packaging material. By thermal dissociation they create reactive spots in the polymer, making the material very liable to form peroxide with atmospheric oxygen, which degrades the 'plastic' properties. In the case of PVC, hydrochloric acid gas (HCl) is generated during the production of the packaging material (forming of bottles and films). This corrosive acid adds to the deterioration of the resin. As HCl can be formed only from chlorinated plastics, the retardant additives are called 'heat' stabilizers, as opposed to the 'antioxidants' used to neutralize the peroxide derivatives and to avoid their formation. Thus stabilizers and antioxidants are always present in plastic resins: their presence in packages for warehouses and supermarkets could be harmful as the light from the neon tubes used in the display halls, has sufficient energy to induce the above-mentioned dissociation. The percentage of most stabilizers is around 0.1% in the final packaging material. Table 6.7 lists some complex phenolic compounds and tin derivatives bonding with sulphur (mercaptans) which are also used. Of course tin compounds are generally toxic; however, they have the advantage of being light-sensitive so that breakdown substances are formed which do not protect the plastic any longer because of their limited life; after use the plastic will be degraded faster by weathering. UV absorbers do not neutralize the peroxides but have the property of absorbing UV-light, like window glass which is transparent to daylight but impermeable to UV-rays.

3. *Technical aids*: in the literature these additives are generally called 'production aids'. They will indeed act as 'aids' during the various steps from raw material processing to final packaging: lubricants to lower the viscosity during warming-up, 'anti-block' aids to make the film surfaces less likely to stick

together, anti-adhesive agents enabling an easy removal of shaped materials from their moulds, anti-static agents limiting the generation of electrostatic charges (friction electricity). Accumulation of static charges can either damage the material or start a fire if sparks fly, or spoil the material because dust particles will be electrostatically attracted.

Anti-fog agents reduce the formation of water droplets on the internal side of packaging films: packages become misted and products look less attractive; these additives remain mainly at the packaging surface where they lower the surface tension of the water droplets so that water itself spreads out in a thin film.

Anchoring agents provide the adherence between polymers and metals e.g. aluminium in composite packages (laminates). The art consists of saturating the polymer surface with molecules which are partly imprisoned in the polymer and partly stick out of its surface: the latter ensure the liaison either by electric attractive power or by chemical reaction.

Chelating agents mainly fix the metallic ions existing as impurities in the polymer.

Finally, we consider categories of additives used for special purposes:

1. Foaming agents: gaseous substances (e.g. freons) now replaced by carbon dioxide liberated during the plastic preparation process which promote swelling of the material. In the final product, the gas is imprisoned in small cells and improves the insulation and sound dampening properties; moreover the small cells full of gas also absorb mechanical shocks (cf. polystyrene foam used to package expensive and fragile goods).
2. Flame retardants: these reduce the flammability of plastics. This aspect is less important for packagings than for the manufacture of upholstery (synthetic leather in cars, synthetic materials for seats in cars and airplanes, etc.).
3. Agents modifying the optical properties of plastic resins: these give the packaging a good brilliancy, hence a greater part of the light is reflected (e.g. plastic surfaces metallized with aluminium).

Table 6.7 Chemical structure of some additives used in plastics for food contact

Trade name	Max. migration mg/6 dm^2	Chemical name	Chemical structure
IONOX 330 (a)	*	1,3,5-trimethyl-2,4,6-tris (3,5-di-tert.butyl-4-hydroxy-fenyl)benzene	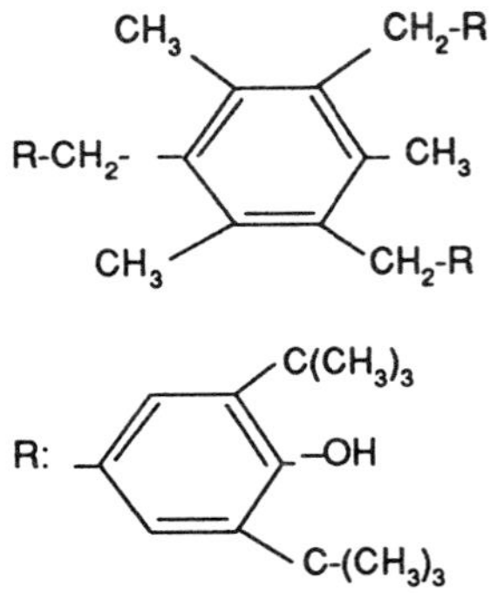
BHT (a)	*	2,6-di-tert.butyl-4-methylfenol	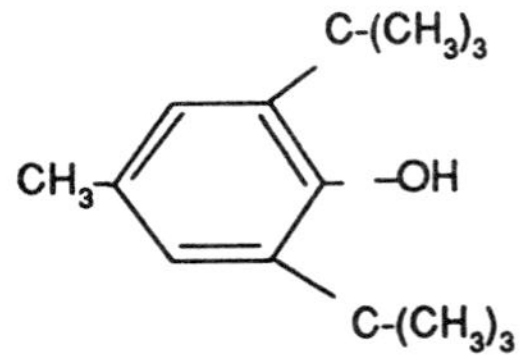

IRGASTAB (a)	1mg Sn	di-n-octyl-tin-di-thio-glycolzuur, 2-ethyl n-hexyl ester	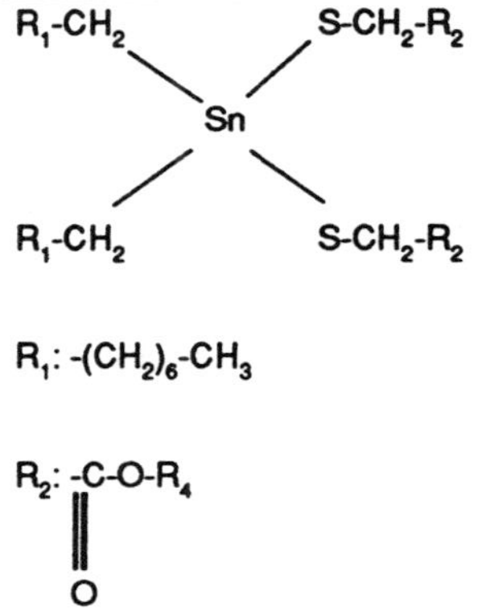
IRGANOX 1010 (a)	*	tetrakis [methyleen(3,5-di-tert. butyl-4-hydrox-fenyl) dihydrocinnamaat] methane	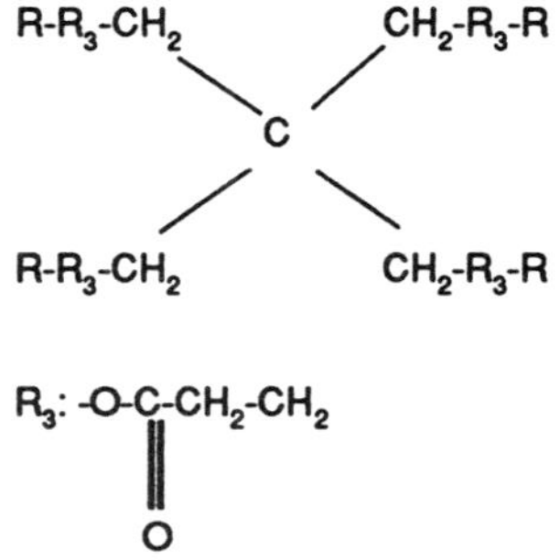

IRGANOX 259 (a)	10	1-6-hexaandiol-bis-3 (3,5-di-tert.butyl-4-hydrox-fenyl) propionate	$R\text{-}R_3\text{-}(CH_2)_6\text{-}R_3\text{-}R$
DEHP (b)	40	di-(2-ethyl-n-hexyl) ftalate	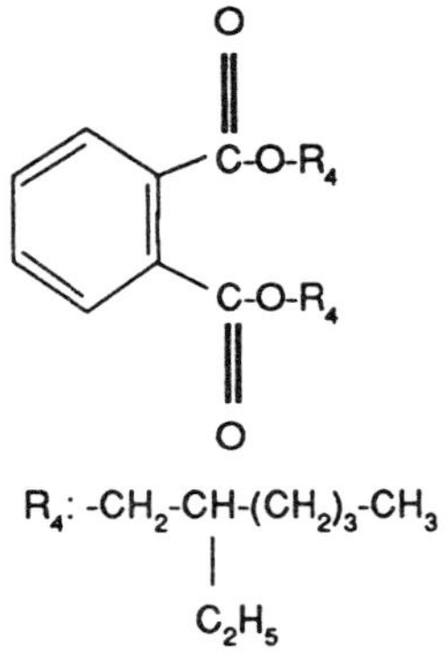
TINUVIN (c)	*	2-(2'hydroxyl-5'-methylfenyl)-benzo triazol	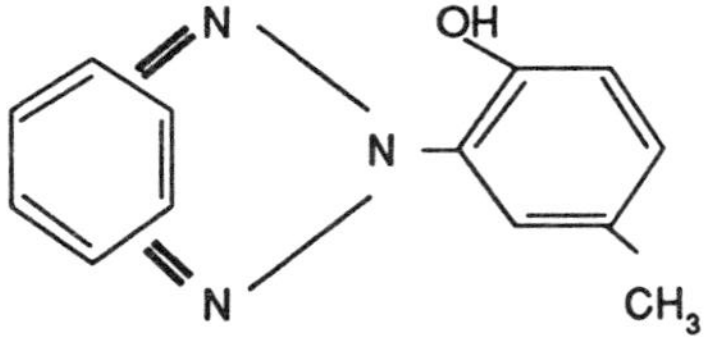

– mercaptanes	0.4 as [SH]	R'-S-H
– peroxides and derivatives	0.05 as[O]	R''-O-O-R'
– phenols and derivatives	0.05 as phenol	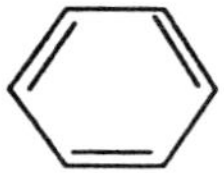 -OH

* presence in the material not higher than 0.1%; (a): stabilizer (antioxidant); (b): plasticizer; (c): UV light absorber

Other compounds such as sulphonamide derivatives from 4-naphthyltriazolyl-stilbene will reflect UV with light of another frequency than that of the incident light creating special light effects (see Table 6.8). These substances can of course be classified in the 'positive list' of additives issued in the RD dated 25.08.1976 *et seq.* In the case of production aids, their concentration in plastic is generally very low, approximately 0.1%; however all are not evenly distributed in the plastic mass because most are active at the surface. This means that as soon as the package is used, they will pass almost completely into the contact medium.

4. *Non-intentional additives*: under this heading we include substances which are neither used nor added during packaging manufacture but which can be generated during the processing by the existing compounds, additives or impurities.

Table 6.8 Production aids of plastics versus relevant chemicals

Type of additive	Chemicals and/or typical representative compounds
Lubricants	Molybdenum sulphite, graphite, teflon powder, amides of fatty acids, esters of fatty acids,potassium-, sodium-, calcium-, zinc stearates
Anti-block agents	Fine silicium oxide, bentonite, clay
Anti-adhesive aids	Hydrocarbons, oils
Anti-static and anti-fog agents	Ethanol and other amines, amides, alcohols, quaternary ammonium salts, synthetic detergents namely sulphonic acids and alkylaryl sulphonic acids, phosphoric acid derivates, polyethylene glycol derivatives
Tackiness agents	Fatty acids (e.g. stearic acid), salicyl acid, chlorinated compounds, piperidine derivatives, titanium
Chelating agents	Triphenyl- and trioctylphosphates, amines, oxamides

This short enumeration of the possible migrants might lead to the conclusion that every packaging material in general, and metals and plastics in particular, are not suitable to package any product. It is necessary to know the properties of the product and of the packaging material to be able to devise the 'best' packaging. Hence it is obvious that plastics either alone, combined with other materials, or used as a coating, can never be considered as single 'pure' materials. When in contact with liquids, migration will occur in most cases; but with solid products, adequate packaging resulting in no migration can be achieved. Table 6.9 lists some plastic packages commonly used for foodstuffs; several experiments have shown (51, 12, 18, 21, 23, 34, 39, 46, 83, 108, 109, 110) that for most 'plastics' the extractive power of oils is greater than that of alcoholic drinks, acids and aqueous solutions. By means of additives marked with radioactive ^{14}C, Figge (50) has succeeded in proving that almost every household vegetable oil seeps to some degree into the packaging (adsorption by the contact surface) and that their behaviour is similar as far as additive extraction is concerned; hence an evaluation of the quality can be based on one particular oil (generally sunflower oil). The fact is that the nature of liquid products can limit the use of plastic resins, especially when the absence of migrants is important (some pharmaceuticals, foodstuffs, chemicals (AR-quality), etc.).

d) Main parameters influencing the migration of packaging components

If we wish to quantify the migration of components from the packaging surface into the packaged product in order to estimate the degradation and/or the potential toxicity, it is necessary to know the factors inducing this transfer. In practice, a fixed weight or volume from a product will come in contact with a corresponding packaging surface during a certain preservation period: the temperature conditions during filling, sterilization and use can be quite different; moreover goods pre-packaged at room temperature are often stored for a few months in the refrigerator or the deep-freeze; some can be reheated in the oven before use.

In brief, we can say that packaging is characterized by the following parameters: the nature of the packaging material and of the contact medium, the contact area and the packaging volume, the contact period and the temperature. As, in principle, a plastic packaging surface, unlike that of glass or metal, can contain a great number of various components (cf. Table

Table 6.9 Plastic resins used for the packaging of some foodstuffs

Product	Plastic resin
Sugar	LDPE
Split peas	LDPE
Meat	PVC; PA/PE; PA/EVA-ionomer; LDPE/EVA
Butter	LDPE/Al
Margarine	PS/LDPE/PS; PP; PVC
Cereals	HDPE/EVA; HDPE/EVOH/EVA
Bread	LDPE/Cellophane; PP/EVA; PP/Cellophane/EVOH
Hard cheeses	LDPE/PVA (coating); PP/PE; PVC; PVDC; EVA/PVDC/EVA PA/PE; PA/PVDC/LDPE-ionomer
Cottage cheese	PS
Dehydrated soups	LDPE/Al/LDPE/Paper/LDPE
Yoghurt	PS
	PE/paperboard/PE
Sterilized milk	HDPE
Pasteurized milk	LDPE
	LDPE/paperboard/LDPE
UHT-milk, Fruit juices	LDPE/Al/LDPE/paperboard/LDPE
Oils	PVC
Mineral water	PVC
Carbonated drinks	PET

6.9) and as the exchange phenomena do not result from chemical reaction although they are mainly determined by some physical laws, we will pay particular attention to them. Column 1 of Figure 6.2 illustrates the most probable distribution of representatives of the already mentioned additive groups as a function of the packaging thickness. Technical aids (A) are mainly present at the surface, plasticizers (C) are distributed almost evenly in the whole mass; conversely, antioxidants (B) as they have been in contact with atmospheric oxygen during the manufacturing thermal treatments and the cooling, have already been partly 'used' so that the concentration inside the wall is

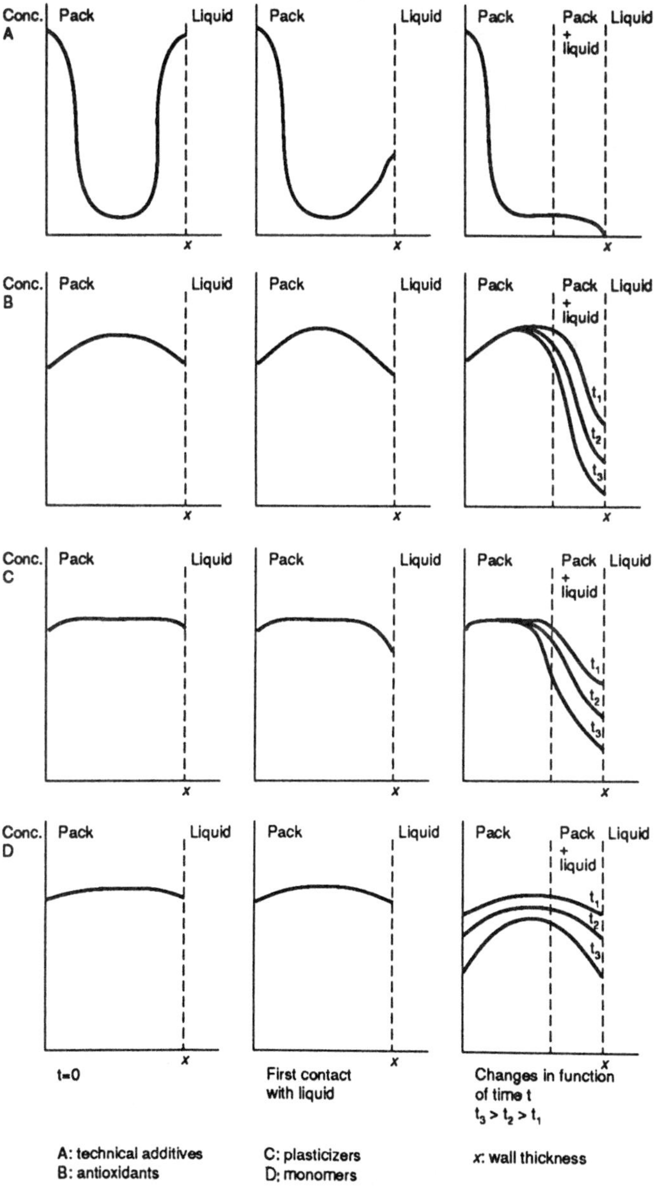

Figure 6.2 Migration of various components from plastics

higher than that at its surface (75, 113). Because of potential diffusion, the monomer concentration near the surface will be lower than inside the wall. Column 2 shows that technical aids are liberated almost immediately at the first contact with a liquid; hence they can be very quickly removed. Column 3 illustrates the penetration of the contact liquid which will involve a further removal (extraction) of the existing additives. This extraction will be particularly important for the antioxidants, plasticizers and non-intentional additives. As already mentioned, gaseous monomer molecules migrate independently of contact with a liquid.

Migration M of a monomer is easy to describe physically and can be calculated on the basis of Fick's laws. A first law applicable to most diffusion phenomena states that the mass F diffusing per unit of time and per unit of surface, is in inverse proportion to the rate of change in concentration C of the component in question as a function of the thickness x. D is the diffusion constant; because the concentration C decreases with time, the equation is negative:

$$F = -D\frac{dC}{dx}$$

The second law describes the potential variation of the concentration of the diffusable component as a function of time:

$$\frac{dC}{dt} = \frac{d^2C}{dx^2}$$

From these two equations, it appears that the quantity m_t liberated by the surface at the instant t (at least at the beginning of the migration) can be calculated from:

$$m_t = m_\infty \ .2 \ \frac{S}{V}\sqrt{\frac{Dt}{\pi}}$$

where m_∞ = available quantity of migrant in question
S = area of the packaging (one side)
v = volume of the packaging material
(i.e. S multiplied by thickness) (50, 105)

In this formula, m_∞ will be considered as constant, i.e. the quantity liberated in the long term. It is important to notice that, according to the t relationship, the quantity liberated at constant temperature after one unit of time (e.g. 1 day) will be doubled only after four units of time (4 days).

From various experimentations (105, 50, 51, 56), it seems this relationship, valid at the beginning of a pure diffusion phenomenon relative to simple gaseous components, is also applicable to all specific migration phenomena especially when 'dependent' migration is preponderant (plastic in contact with liquids such as water, oil, etc.)

The evolution of the 'global' migration, as the sum of 'specific' migrations, can also be represented by the above mentioned $\sqrt{t}$ relationship (cf. Figure 6.1).

$$M_t = M_\infty \,.2\, \frac{S}{V} \sqrt{\frac{Dt}{\pi}}$$

where M_t = total quantity liberated by the packaging surface after a contact period
M_∞ = total quantity of additive liberated in the long term
D = global diffusion constant

From these equations we conclude that the global migration M_t of plastic packaging components:

1. Is in direct proportion to the total quantity of additives (M_∞) existing in the package and available for removal.
2. Increases when the contact period is longer. At the beginning the migration develops rather quickly, afterwards it will increase slowly as the contact continues. It tends to a

maximum value which depends very much on the temperature (cf. Figure 6.1).

3. Is in proportion to the packaging area. However there is no constant ratio between the area and the capacity of the package: this ratio becomes smaller as the volume increases. This means that migration is relatively lower in the case of large packages (e.g. 1.5 litre) than for smaller ones providing the same capacity (e.g. 3 × 0.5 litre).

Generally we can say that the quantity of migrant liberated per unit of surface is independent of the content volume. However, as the total packaging surface increases more slowly than the volume enclosed, the concentration of migrant will always be lower in the case of big packages than for small packages for the same capacity of volume. The following simple calculation of, respectively, the surface and the volume of a cylindrical package with radius R and height H, demonstrates this:

- lateral surface area : $2 \pi R H$
  top and bottom surface area : $2 \pi R^2$
  total area : $2 \pi R (H + R)$
- volume : $\pi R^2 H$
- $$\frac{\text{surface}}{\text{volume}} : \frac{2 \pi R (H+R)}{\pi R^2 H} = \frac{2 (H + R)}{RH} = 2 \left(\frac{1}{R} + \frac{1}{H}\right)$$

Hence this ratio decreases progressively when bigger volumes are taken into account (higher H with R constant or greater R for the same H) or, conversely, increases when smaller volumes are considered. Therefore, migration values are expressed in weight of migrants (mg) per unit of surface (dm^2). As a 1 litre cubical package (packaged product about 1 kg) represents an area of 6 dm^2, it is possible, when the contact area is difficult to calculate, to express migration values in mg per kg of packaged product provided the result is multiplied by 6. Practically a migration value of 10 mg/dm^2 is thus equal to 60 mg/kg.

4. Global migration also increases when the temperature rises.

Of course, the 'temperature' factor does not appear in the above equations but the diffusion constant (D) and the migratable quantity of additives (M_∞) can vary considerably with the temperature. The influence of temperature has been described by different formulae in the literature: in any case the migration seems to increase with temperature not in a linear but in an exponential manner (74). This can also be noticed in Figure 6.1 where specific migration rates at various temperatures are shown for one particular plastic. Generally the maximum migration (specific and global) will be achieved quite quickly and repetitively at 40°C (33). This is one of the reasons why standards fix 40°C for migration tests (cf. Table 6.10, 10 days–40°C test).

Table 6.10 Simulation conditions for migration determination

Actual conditions		Simulated conditions	
Period	*Temperature*	*Period*	*Temperature*
a few months	<5°C	10 days	5°C
a few months	<30°C (RT)	10 days	40°C
shorter periods	<30°C (RT)	2 hours	40°C
shorter periods	<70°C (HF)	2 hours	70°C
shorter periods	<100°C (past)	30 minutes	100°C
shorter periods	>100°C (ster)	30 minutes	120°C
shorter periods	>100°C(boiling)	2 hours	150°C
shorter periods	>100°C(baking)	2 hours	175°C

RT: room temperature; HF: hot filling; past: pasteurization; ster: sterilization
[Source: RD dated 25.07.1976, Belgian Official Journal 24.09.76]

5. Migration also depends on the kind of packaging and on the nature of the liquid in contact with the packaging surface. These aspects have already been detailed earlier and these parameters are mentioned here for completeness.

Factors relative to plastic packaging, listed systematically here, also apply (no. 1 excepted) to other materials. The exchanges between packaged content and packaging surface always increase with temperature, contact period and area; in these cases too the remarks made about the ratio area/capacity are applicable. Finally, we must note that migration is positively induced by mechanical stresses (vibrations occurring during transport) to which packages are exposed, and by exposure to ionizing radiation.

e) Determination of migration, relevant legislation and estimation of packaging toxicity

To determine the specific and/or global migration, in order to appreciate the quality of packaging products, the contact conditions existing in practice should be reproduced as accurately as possible: for practical reasons, long preservation periods are shortened by performing the tests at higher temperatures. So, a preservation period of a few months at room temperature (maximum 25°C) is simulated by a 10 days–40°C test, preservation in deep-freezer or refrigerator by 10 days–5°C etc. (cf. Table 6.10). For the research and determination of small quantities of known and unknown migrants, the contact liquid cannot easily be the same as in practice, mainly because food is generally of a complex composition and the determination of non-intentional additives migrating from the packaging would be an impossible task. This is the reason why test liquids are used to simulate four categories of foodstuffs: namely neutral aqueous, acid, alcoholic and oily foodstuffs which are represented respectively by water, acetic acid-water, alcohol-water and refined olive oil or sunflower oil. The simulation conditions and liquids prescribed in the RD dated 25.08.1976 and 25.04.1980 are listed in Tables 6.10 and 6.11.

Table 6.11 Foodstuff categories with the relevant simulation liquids

Foodstuffs	Simulants
Liquid non-acid foodstuffs and drinks (pH >5)	Water distilled twice
Acetic acid	Acetic acid of same concentration in water
Liquid acid foodstuffs and drinks (pH <5)	3% (weight/volume) acetic acid in water
Oils and fats	Refined olive oil or sunflower oil
Alcoholic drinks (alcohol content >15%)	Ethanol at the same concentration in water
Foodstuff with alcohol content >1%	15% (volume/volume) ethanol in water

[Source: RD dated 25.04.1980, Belgian Official Journal dated 18.07.1980]

For the determination of the actual migration a precise area of packaging material, either one or both sides, placed in specially designed cells, is exposed to the simulation liquids and conditions described earlier. The ratio between the packaging surface area (in cm^2) and the volume (in ml) of the test liquid ranges from minimum 0.1 to maximum 2.

To determine the specific migration, the concentration of the migrants in question (e.g. tin, lead, plasticizer, antioxidant) in the test liquid is determined according to an appropriate analytical method.

In principle, two methods may be used for the determination of global migration: on the one hand, measuring the weight loss of the packaging material and, on the other hand, determining the residue in the simulation liquid after evaporation. Results of both methods are almost never the same: when checking the packaging weight loss, a potential increase in weight due to absorption of liquid (18, 39) or to eventual reactions (87, 108) should be taken into account. When measuring the evaporation residue, a potential loss during the evaporation should be taken into account. However, practice has proved that results obtained by evaporation of the simulation liquid are better able to be reproduced; hence it is used for all simulants, olive and sunflower oils excepted. Tests with oils face many problems as, after exposure, packages must be treated with a scouring agent in order to eliminate the olive or sunflower oil infiltrated into the packaging surface. This treatment could eliminate other additives too, so that inaccuracies can be induced. Measured migration rates are believed to be generally lower than the actual ones. On the basis of these data, migration rates can be worked out. The RD dated 25.04.1980 includes a table listing all foodstuffs in the same order as in the relevant Belgian legislation (Chapter III) with an indication of the liquids prescribed for the tests; notice that a reduction factor of 3 or 5 is given especially for foodstuffs containing some fat (meat, cheese, cookies, etc.); this means the final result of the migration test should be divided by this factor (3 or 5).

According to the R.D. dated 25.08.1976, the global migration must be lower than 10 mg per (dm^2) of packaging surface, or lower than 60 mg per kg of packaged product. The applicable reduction coefficients indicate that for some products containing oil or fat, the simulation procedures are more severe than the real conditions of contact. So solid fats (e.g. margarine) are less able to induce migration than the liquid simulant, sunflower oil. Each simulation test requires that the actual conditions are approximated: opinions are divided as to whether the specified test liquids always reproduce the actual phenomena: so, in the USA the Food and Drug Administration (FDA) has chosen n-heptane as the simulation liquid for oils and fats (60, 33, 21), and in Switzerland pentane has been chosen. Extended migration studies (N. Kruyf from TNO .....) show that iso-octane can be used for most simulations. We think that an aqueous solution containing 3% n-heptane in an accelerated migration test, in which the heptane is emulsified in fine droplets in the water, has a behaviour very similar to that of pure n-heptane. Hence we may assume such solutions could better simulate milk, as its butter fat (± 3.5%) is also present as small droplets (18, 105, 107, 108, 109).

These data lead to the conclusion that 'migration' is a topical problem and remains the subject of many research projects. As already explained, the maximum global migration is fixed by RD; the measured global migration serves only as an indicative value of importance but does not give direct indications of the toxicity or the safety class of packages, as a complex mixture of additives and/or components can pass into foods. Every resource used to manufacture materials as well as to process them in order to produce packages, must be mentioned in 'positive lists' in the Belgian and European legislation: no other substances are allowed. With each application to use a new reactant or additive, results of toxicity tests among other things must be included, although this proves nothing about the safety of the assembly; it only limits the potential harm. If the specific migration rates of the packaging components as well as the total surface in contact with the foodstuffs are known, it is possible to calculate the average quantity of additives ingested daily with food, taking into account the habits and routines of life. Should such 'daily dose' of each potential component be lower than its ADI, we could not conclude that the dose of the mixture of these components, as regards the global migration, would be innocuous. First, all the components migrating are not identified, and moreover it might happen that, ingested over a long period with the daily food intake, these components could accumulate in the body; hence pathological troubles might be induced as soon as a limit value is overstepped. The potential long-term effects are not yet known! Moreover the effects of all 'foreign components' already existing in our food before packaging must be taken into account (115).

Tests able to correlate all influences of packaging and food additives on the body are still not defined. Of course, the literature describes tests judging packaging materials on the basis of the reactions they provoke when inoculated into laboratory animals (11). According to Dr Autian and his collaborators, the 20% minimum slowing-down observed in the cell growth of simple fibroblasts is caused by packaging migrants, an important datum that could be used to determine the toxicity of packaging materials for human beings (11). Requiring the 'no presence level' for harmful components is nowadays still not technically feasible; moreover it is not always necessary as, according to Gilbert, zero migration can be obtained by keeping the concentration of additives in the plastic below a certain value (61). Indisputably, in the future the results of various tests on packaging migrants (determination of mutagenicity, teratogenecity, carcinogenicity, etc.) will be taken into account in estimating the reliability of materials and wares designed to be in contact with foods. In this respect, qualitative tests have already been performed by the Belgian Institute for Commodity Science and Technology (BIWT-IBTM) in collaboration with the University of Gent (RUG) and the Free University of Brussels (VUB).

These arguments should not lead us to the conclusion that packaging, especially metal or plastic, is not safe: we have already considered many cases ensuring zero migration. However, we could conclude that, as far as migrations are concerned, the safety degree is now tending to be determined or guaranteed. Nevertheless, vigilance is recommended; especially when new technologies and/or food preservation methods are introduced (e.g. microwave oven, ionizing radiation, etc.).

Legislation relating to packaging has already been detailed earlier. A list of the main EC directives is attached; generally the national decrees follow the standards of the EC in principle, the simulation conditions and the maximum allowable migrations prescribed in Belgium are the same as those fixed by the EC and most commercial partners outside Europe, and in many respects are equivalent to those of the FDA. The harmonization in Europe is planned for 1 January 1993; the national testing laboratories will also be recognized and accredited by the EC member-states.

The evolution of Belgian legislation relative to packaging during the last 19 years is worth mentioning. In the law dated 20 June 1964, there is only one paragraph concerning the 'reliability of packagings', i.e. Art. 2.2 stating 'the King is empowered to determine the use of wares and materials designed to be in contact with foodstuffs, their control and their interdiction if their use is or seems to be dangerous or harmful'. Art 3.2 calls 'technological' the additives coming from the packaging. In more recent decrees, attention is mainly drawn to the limitation of migrants and additives (cf. positive lists in the RDs dated

1974 and 1976), on maximum allowable residual values in the packaging (mostly 0.1%) and on relevant determination methods. A further step in this direction is shown by the decrees mainly related to labelling, which also insist on many aspects of the packaged content.

Note also in the evolution of the decrees that more and more importance is given to 'Man and his environment': hence the 'safety' of the marketing and use of products is more directly linked to the results of various tests studying, for instance, toxicity as a function of bio-accumulation, the influence of allergies, of heredity, etc. (RD dated 24.05.1982). This shows also clearly that legislation – limited by technological possibilities and progress although following them – determines stricter quality criteria for packaging in general and foods in particular.

4 PRINCIPLES OF 'TAMPER-EVIDENT' AND 'CHILD RESISTANT' CLOSURES

a) General considerations

We introduced in the definition of packaging the importance of the presence of a seal to distinguish packagings from wrappings. Seals protect manufacturers as well as consumers: manufacturers want to deliver their products unspoiled and with an assurance of the quality that existed at the end of the production process. Packaging and its closure, provided with a seal, must give the consumer the assurance that he has received a product quality as designed by the manufacturer. Of course, the manufacturer remains responsible for his product until the moment of opening. Tamper-evident closures aim to avoid involuntary or maliciously intended spoilage or any form of uptake of the content; the system must give evidence that packages remained closed during their life-span, from manufacturing to the moment of consumption. Well known are the early closures where a waxed paper-wrap was insured by sealing wax. Packages can be made tamper-evident by many methods: sacks, pots and water-proof-packages are closed by joining plastic parts under heating and pressure; the thermoplastics will melt and combine, and a tight closure is likely to be obtained after cooling.

A version of this system is in the closures provided with a tear-strip (cf. the Tracer-safe strips of Johnson and Jorgensen) and those provided with a metallic or plastic ring which will break during the first attempt to open the package.

Many products, particularly pharmaceuticals, are distributed in blister packs. A double function is aimed for here: tamper-evidence and, where required, indication that doses have been properly used.

b) Criteria for child resistant closures (CRCs)

Hermetic closures are not always sufficient to guarantee some safety requirements. Indeed, the Belgian Antipoison Centre, for example, receives nearly 50 000 calls yearly to treat and to advise about injuries caused by accidental usage of packaged products such as dangerous goods, pharmaceuticals, detergents, etc. More than 50% of the victims are children who gained access to the products in question. Therefore, there is a worldwide need to develop regulations to safeguard children. Closures have been produced to reduce the ease of opening in order to avoid abuse or misuse of products. An international standard ISO 8317 (1989) is already in force, but control will become more stringent in the coming years.

The design of CRCs is faced with a controversy: most systems are based upon the requirement to apply two actions simultaneously, combining handling in conjunction with the traditional movements of turning or lifting; so the physical skills for opening the closure are higher than those possessed by most children below the age of three. The squeezing force of children ranges from 0.15 kgf/m^2 to 0.36 kgf/m^2 respectively when one or both hands are used. The force they can exert is as much as three times higher when they use their mouths. On the other hand, elderly people should not be prevented from access to wanted or needed products (e.g. medication). Most common closures are those based on combined and subsequent handlings which children cannot perform without training. Some child resistant closures are illustrated in Figure 6.3.

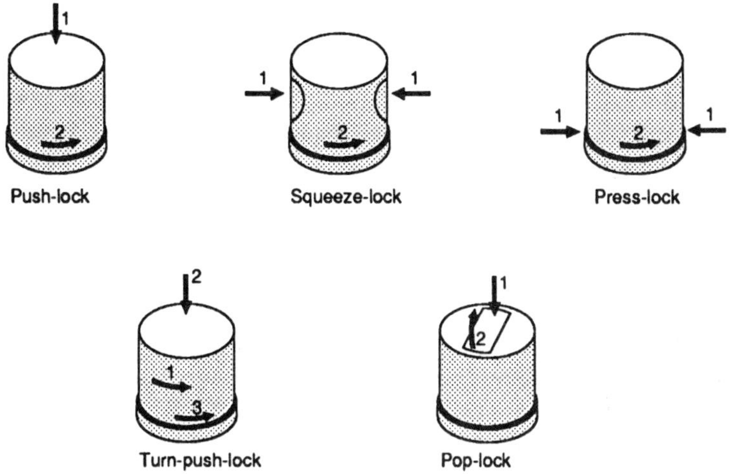

Figure 6.3 Child resistant closures

The CRCs must achieve the purpose for which they are intended: protection! Their efficiency is tested with groups of children and of adults. In the case of children, the following criteria must be fulfilled in order to consider the closure 'child resistant':

- 85% of a group of 200 children must not succeed in opening the closure without demonstration during a test period of five minutes.
- 80% of the group must not succeed after demonstration.
- 80% of the children must not receive access to an amount of the product as to cause serious injuries or damages.

The test group consists of children aged between 42 and 51 months.

In the case of adults, there are normal procedures whereby age, number of people and the ratio of men to women are prescribed. For adults from 18 to 45 years, for example, a test group of 70 persons is considered and the following criterion must be fulfilled:

- 90% of the people must achieve access to the content after reading the instructions.

The effectiveness of child-resistant closures can be calculated from the test results or can be visualized in a graph. To calculate the security level S (in %) of a closure, the following relationship can be used:

$$S = \frac{X - x}{X} \cdot 100$$

where S = calculated security factor (in %)
X = number of test persons
x = number of failures

and $S \geq N$ where N= norm-limit (90, 85, 80%)

Some packages meet both the functions of tamper-evidence and child resistance at the same time: well known are the 'Single Unit Dose Packing', the 'Original Pack Dispensing' and the Aluminium-faced blister pack: when opened, the closure cannot be re-installed. These types of closures are mostly used for packaging drugs: only the fact that each unit is packed separately reduces the risk of misuse and/or intake of high doses.

It is clear that technology cannot offer solutions to all sorts of problems or provide protection to all possible risks with which children can be faced. There are limits stated by realistic occurrence of facts as shown and treated at the beginning of Chapter VI. As in many cases where health and safety are concerned, there is the important rule of 'prevention instead of healing'. In a society where all sorts of risks become interrelated, a chief learning task is undoubtedly the responsibility of parents and educators. The role of legislators must be considered as a helping hand to encourage the prevention of children being exposed to dangerous situations. And here again, a justified choice of packaging can be helpful.

VII
Justification for packaging

1 INTRODUCTION

Good housekeeping starts by holding expenditure in balance with income, and this reasoning can be extended to groups, societies and industrial activities as well. Balances are based upon realistic, economic and concurrent considerations.

In the advance of packaging, we recognize its primordial role in the fight against shortage, spoilage and loss of food and goods. As shown before, packaging offers better fulfilment of wishes and requirements, a safer distribution of products and can be very helpful in the fight against hunger worldwide.

Scientific research has led to a better understanding of all the parameters governing product deterioration and to the manufacture of materials with properties as specified on the drawing board. The consequences of production schemes and their scaling-up have already been discussed in earlier chapters.

The actual interest in treating the 'environment' as a valuable and thus as an economic resource, originated in the early 1960s: people became aware of the finiteness of the world and of the restriction of resources of materials and energy. Production became analysed by considering the subsequent energy needs. At the World Energy Conference held in 1963, Harold Smith presented for the first time calculations with respect to cumulative requirements of energy for the production of chemical intermediates. Afterwards, several studies were published in *The Limits to Growth* (Meadows) and in *A Blueprint for Survival* (Club of Rome). The predictions of dramatic consequences from an increasing world population and goods production upon the exhaustion of resources (materials and fossil fuels), and the climatological changes due to the thermal load on the atmosphere and surface waters, gave rise to a review of industrial processing with regard to environmental impact. Coca Cola commissioned the Midwest Research Institute in 1969 to carry out systematic evaluation of packaging materials for drinks with respect to their contribution to environmental pollution and exhaustion of energy sources. The awareness of the finite capacity of ecological assimilation of pollutants became acute: the 'ecobalance' was born and the method of its quantification was known in the USA as REPA 'Resource and Environmental Profile Analysis'. Since 1988 the REPA-studies have been extended to analyse the methods of solid waste disposal as well. As a result, all the possible interactions encountered in a product's life-span are studied from 'cradle to grave'. Public opinion is focused on packaging in general (30% of household waste) and plastic packaging in particular (7% of household waste); the disposal of solid household waste is an acute problem, and each consumer is confronted daily with the disposal of his own voluminous post-consumer packaging. In these claims, the advantages and conveniences that modern packaging offers tend to be forgotten. Solutions must be found, undoubtedly, and here a fundamental study of ecobalances can offer important information to manufacturers, consumers and officials for establishing directives and guiding actions in order to prevent pollution, to reclaim materials and energy and to optimize disposal systems.

2 PRINCIPLES OF ECOBALANCE, ECOPROFILE AND LIFE-CYCLE ANALYSIS (LCA)

The evaluation of the impact that production and use of products have upon the environment are generally delineated in the literature by the terms 'ecobalance', 'ecoprofile' or 'life-cycle analysis'. Although the intention or the purpose of these studies can be the same, we define, as far as packaging is concerned, each aspect differently by principle.

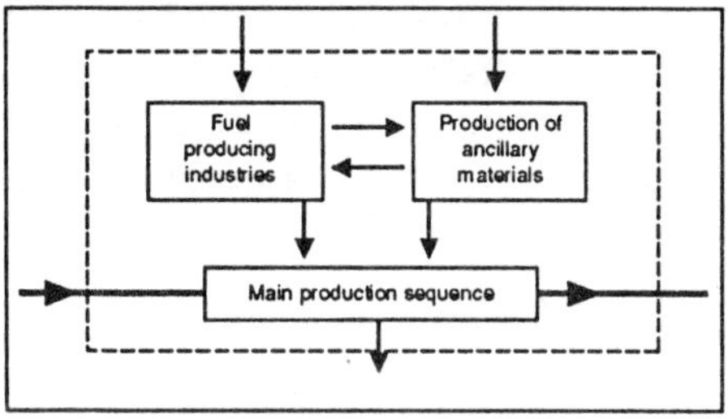

Figure 7.1 Model of packaging production

Ecobalances can be regarded as the evaluation of the environmental impact of each process, part of the whole subsequent series of handlings by which packages are made and treated. So we consider the ecobalance of energy production, the production of materials, transportation, distribution, waste treatment, etc. Ecobalances can be expressed as the energy (in MJ) and the materials required, and the emissions produced per unit of packaging material (in kg) or delivered number of packages (per thousand). Packaging production can be evaluated by a step-by-step analysis following the model pictured in Figure 7.1.

The life-cycle analysis can be regarded as the summation of all ecobalances encountered in the material flow to produce packages and their waste treatment. The pattern to evaluate and calculate a 'life-cycle analysis' is shown in Figure 7.2.

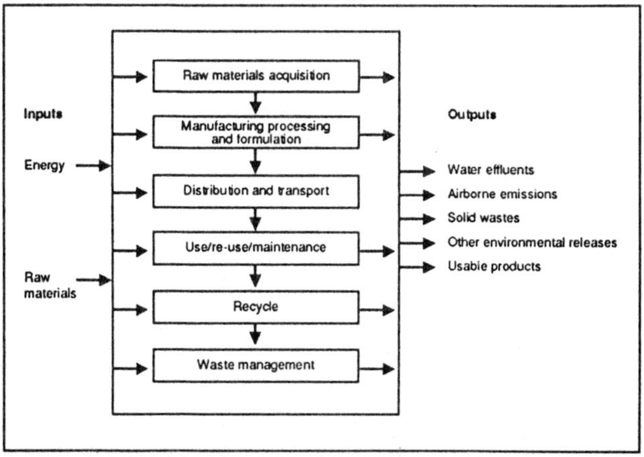

Figure 7.2 Evaluation of an LCA

Here we consider the environment as a donor of all sorts of resources (materials, oxygen, energy, water, etc.) and as an acceptor of the products and all types of waste. Ecobalances and life-cycle analyses are mainly based upon considerations of the consumption of materials and energy and the associated production of materials and wastes.

The published figures and analyses are not coherent. There are many reasons for this; an exhaustive list goes beyond the purpose of this chapter, but it is interesting to look at some bottlenecks and possibilities which can distort or soften the reality. Differences originate in:

- the access to resources and the distance between mining and processing – these differ from one country to another,
- the ways in which electricity and/or forms of mechanical energy are produced. Electricity from water power is likely to be different from an environmental point of view than electricity from heat either produced by combustion of fuels or from nuclear fission reactions,
- the technology of processing. The extraction of sulphur, e.g. from fuels or the treatment of combustion gases to absorb sulphur dioxide enhances other waste-streams and energy uses,
- the description and the delineation of the system boundary considered can be different. The energy content of waste can be taken into account as a negative energy content (energy which can be recovered as heat and/or as electricity [1/4 of heat]); recycling, re-use or reprocessing (e.g. remelting of plastic scrap or of collected post-consumer plastic materials) are different. So the production processes of materials include waste disposal or recycling ('from cradle to grave' or 'from conception to resurrection'),
- the LCA depends upon the size of delivered product: e.g. sugar at a coffee shop (5 g unit size); sugar-bag of 1 kg (household package); sugar-sack of 5 kg.

The units must be carefully understood, otherwise quite different conclusions can be formulated. Let us consider, for example, some energy needs for packaging expressed per thousand packages or per unit of volume of packed product as presented in the table below:

Material/ package	Volume (in ml)	Energy per 1000 packs (in MJ real case)	Energy calculated per 1000 litres delivered (in MJ)
Brick/pack	587	3.700	6.302
Glass/bottle	502	3.752	7.475
Tin/can	374	5.774	15.437
Alu/can	372	6.513	17.508
PET/bottle	1511	13.740	9.093

From these data nothing can be deduced from the environmental point of view. We refer here to Tables 4.5 and 4.7 where several kinds of energy forms are indicated.

Estimation of the possible environmental impact of production schemes is given by the so-called eco-profile. The ecological load of an industrial activity is expressed by 'critical values': i.e. the volume of air or water necessary to dilute an amount of product emitted to the atmosphere or released into surface waters. To reach the estimated air or water quality, standards are taken into account (see Tables 2.4 and 2.5). Also, the volume of waste to be finally disposed (compacted material, ash, etc.) is also given by an eco-profile. In Table 7.1 we summarize some data as published by BUWAL (1990).

It is clear that the emission or the release of products depend upon the technology and the proper maintenance and safety of the gas or water treatment systems. The energy requirements are typically dependent on the material. The evaluation of production by the description of critical volumes must be considered as an attempt to discover the weak points in an industrial process which are to be improved.

Critics of this method of evaluation are concerned mainly by the fact that it is not acceptable that the total volume of air or water available on earth will sufficiently dilute the waste. The

Table 7.1 Eco-profile of some packaging materials

Material	Energy needs (Mj/kg)	Critical volume of air (1000m^3/kg)	Critical volume of water (m^3/kg)	Volume of solid waste (cm^3/kg)
Glass				
56.2% recycled	7.5	293.0	1.3	237.9
74.8% recycled	6.8	242.5	1.2	147.2
100.0% recycled	5.9	165.4	1.1	24.2
Paperboard	33.1	381.4	948.3	287.0
Paper (bleached)	47.3	718.7	1487.7	355.2
Corrugated board	16.3	186.1	136.9	246.5
Aluminium (hydro-elect.)				
zero% recycled	126.1	2611.0	638.6	967.6
100% recycled	12.6	300.7	1.9	276.0
(elect. from fossil fuels)				
zero% recycled	171.2	4048.6	640.3	1902.9
100% recycled	15.6	354.4	2.0	281.0
Plastics				
HDPE	46.9	221.2	107.3	278.3
LDPE	47.4	231.2	107.3	292.3
PET	69.5	692.4	119.7	295.1
PVC	42.5	669.0	307.2	401.5
PP	50.0	331.5	122.2	300.9
PS	56.2	675.9	60.2	338.4

[Source: BUWAL; Switzerland (1990)]

environment and the living organisms it contains are full of cleaning and recovery mechanisms; it is well known that persistent pollutants can be concentrated by passing from one organism and then further on through different species in the food-chain, increasing to dangerous and toxic levels (e.g. birds dying from eating contaminated flies, etc.). So we do not know and we cannot predict how contaminated our aquatic food, for example, will become in the future. It is also impossible to estimate the decrease of pollutants by solar radiation or by rainfall, or what influences pollutants will have on life expectancies in the future.

The ecological-impact studies as described above are not specific to packaging but are applicable for each industrial process. The main purpose is to improve activities which might otherwise interfere with the ecological equilibrium, and to avoid unnecessary use of materials and energy.

We may conclude that as far as packaging materials are concerned, these studies should not result in the rejection of some materials and thus the conveniences, facilities or possibilities they offer. It must be possible for the managers and the consumers to make a preferential but justified choice to meet their requirements. Biodegradable plastics, for instance, have their specific field of application: it would be a mistake to ignore them and it could even be worse to replace all plastics by biodegradable ones!

3 SEEKING THE OPTIMUM PACKAGING CHOICE

This section can be considered as the summary of previous chapters: it is addressed to all consumers, to the authorities and producers and users of packaging materials.

Let us start by saying 'there is no perfect package, all include a compromise, and all packagings on the market nowadays are good for their specific intended purpose'. We must bear in mind that each package is the result of research aimed at achieving the safe delivery of a product irrespective of time and distance, furthermore each package is designed to withstand specific stresses (transport, storage, etc.) and to fulfil specific requirements as formulated by individuals, families, institutions (hospitals, schools, restaurants, etc.) or industries.

There is no single material on the market which, from a realistic and economic point of view, can be useful to pack every product (equipment, food, chemicals, pharmaceuticals, dangerous goods, fuels, etc.); the choice would be limited by the size required and, of course, all disposal problems could be easily solved (see Chapter IV). The purpose of the discussion here is to review briefly the most important 'parameters for choice', discussed extensively in earlier chapters. It should be made clear that no preference is imposed here, but each manufacturer and consumer must be able to choose which solution he considers as justified and as the 'best' one to fulfil his wishes, requirements and needs.

The choice between the materials offered today is not easy, because it is fundamentally a multi-criteria analysis of important parameters, the weighting factors of which may be decided differently by each individual. First of all, the amount of product to be packaged, and thus the size of the package to be sold, should be related to the consumption rate or needs during the period that quality can be ensured once the package is stored or opened; we should not, for instance, buy 2.5 litres of sulphuric acid when only 100 ml is used over one year or more. This reasoning is also valid for food: we should not buy a 1 litre container of cream when only 250 ml is consumed within one week – 750 ml will deteriorate and become useless as cream after one week. Furthermore, the loss of 750 ml of cream is worse and costs more than the supplementary amount of energy and materials required to produce four 250 ml bottles.

The aspect of size is very important for general consideration with respect to the environment (portioning goods requires more packaging material and energy than those required in producing one big container with the same total content). However, the size remains theoretically independent from the properties of the packaging material and the product to be packed.

The most valuable parameters determining packaging choice are now considered in order of importance:

a) The product to be packed

The product dictates special features from the package and the packaging material: cushioning, permeability, water- and oil-resistant properties, etc. Not all packaging materials are compatible with all products; there are fundamental restrictions; e.g. alkalis will not be packed into aluminium boxes; nor can sparkling water be packed in conventional PE-bottles, etc., so the choice of packaging material or combinations depends upon the product to be packed.

b) The surroundings

We have discussed extensively the protective function of packaging: it must offer resistance to mechanical stresses encountered during handling, transport, distribution and stacking, to physico-chemical stresses caused by variation of temperature, humidity, gases and vapours, and finally to biological attacks.

So the choice of packaging material and packaging design depends very much on the properties of the environmental stresses to which packaged goods are exposed during their existence.

c) The living standard and way of life

We have demonstrated that packaging bears witness to the advances and the management of science and technology and thus to the innovation, creativity and the living standards of the society that enables their production.

Here people should be able to choose that package (size and material) which fulfills their needs and wishes. The packaging materials (and/or size) for goods to be carried during climbing mountains may be quite different from those encountered for daily use. Goods can be packed in different sorts or combinations of materials, each of which offers special features and convenience by weight and by visual or tactile properties. So the choice of the appropriate material is governed by considering fundamental needs and wishes as encountered in daily use and in recreation, travel, journeys into space, etc.

d) Ecological factors

Once materials have been selected on the quality statements discussed in (a) and (b), the ecological implementations should be taken into account. Here ecobalances and eco-profiles can provide extremely important information. As explained before, this may not be exclusive: on the one hand, there is the responsibility to give the right undistorted information and on the other the duty to consider the data on the basis of realistic understanding of the facts. Ecological impacts of packagings due to their production, use and disposal are difficult to evaluate properly and the synergetic effects of emissions and disposal of various components in the air, water and soil are still under study.

The fair solution to a compromise between the convenience offered by packaging and the problems encountered in managing various types of waste, is beginning to come under control: it is a challenge to improve actual technologies and to promote the search for new materials and applications. Recovery and recycling of materials can be successful, and depend mainly on selection, collection and efficient separation of materials. There is no technology existing for solving all problems; each individual should be aware of his responsibility for the success of the waste prevention and material recovery schemes.

e) Aspects of safety

Packaging safety has been discussed in Chapter VI: there is fair practice in producing and handling packages and in observing the regulations concerned. Migrational behaviour of most materials has been improved during the last two decades, but investigations are still going on for better understanding and control.

New technologies (ionizing radiation, recovery of scraps and wastes) can give rise to new aspects that interfere with quality control. Coating technology has also advanced and opened the market to a great variety of materials with a wide range of applications. Safety control has become a normal reflex when new materials or technologies are introduced.

Safety of packaging is relative: when broken or opened, packages can cause injuries accidentally or by misuse. The correct use of packaging depends undoubtedly on education; the misuse of some package contents can be restricted by the use of child resistant closures.

Several aspects concerning the choice of packaging are worth noting separately but many overlaps are possible. Nowadays, parameter (d) is likely to be used as the banner for all sorts of actions whereby economic profitability and ecological purpose become intrinsically combined and indistinguishable. Optimization of packaging choice nevertheless starts with the statement of the package size.

The properties of products (a) and surrounding factors (b) act in a restrictive way: they cannot be bypassed. The safety (e) and ecological considerations (d) can be regarded as conditional and remain established and under control by decrees and directives. Active cooperation and support by manufacturers, packaging users and the population is of great value in learning from and solving problems encountered.

The final choice may be based upon considerations of safety (e) and the fulfilment of wishes and requirements (c).

Packaging concerns us all; its advance and efficient management carries a price and this is our concern, too.

Appendix 1

LIST OF TABLES

Appendix 2

LIST OF FIGURES

Appendix 3

List of EEC directives for food packaging materials and of legislative developments in Europe concerning waste management

1 Food contact materials

- 76/893 EEC (OJL 340 – 9.12.76); Adoption: 23.11.1976; Application: 26.11.1978
  on the approximation of the laws of the member states relating to materials and articles intended to come into contact with foodstuffs.

- 78/142 EEC (OJL 44 – 15.2.78); Adopt: 30.1.1978; Appl: 26.11.1979
  on the approximation of the laws of the member states relating to materials and articles which contain vinyl chloride monomer and are intended to come into contact with foodstuffs.

- 80/590 EEC (OJL 151 – 19.6.80); Adopt: 9.6.1980; Appl: 1.1.1981
  determining the symbol that may accompany materials and articles intended to come into contact with foodstuffs.

- 80/766 EEC (OJL 219 – 16.8.1980); Adopt: 29.4.1981; Appl: 1.10.1982
  laying down the community method of analysis for the official control of the vinyl chloride monomer level in materials and articles which are intended to come into contact with foodstuffs.

- 81/432 EEC (OJL 167 – 24.6.1981); Adopt: 29.4.1981; Appl: 1.10.1982
  laying down the community method of analysis for the official control of the vinyl chloride monomer level in foodstuffs.

- 82/711 EEC (OJL 297 – 23.10.1982); Adopt: 18.10.1982; Appl: 1.1.1985
  laying down the basic rules necessary for testing migration of the constituents of plastic materials and articles intended to come into contact with foodstuffs.

- 83/229 EEC (JOL 123 – 11.5.1983); Adopt: 25.4.1983; Appl: 1.1.1985
  on the approximation of the laws of the member states relating to materials and articles made of regenerated cellulose film intended to come into contact with foodstuffs.

- 84/500 EEC (JOL 277 – 20.10.1084); Adopt: 15.10.1984; Appl: 17.10.1987
  on the approximation of the laws of the member states relating to ceramic articles intended to come into contact with foodstuffs.

- 85/572 EEC (JOL 372 – 31.12.1985); Adopt: 19.12.1985; Appl: 1.1.1991
  list of simulants for testing migration of constituents of plastic materials and articles intended to come into contact with foodstuffs.

- 86/388 EEC (JOL 228 – 14.8.1986); Adopt: 23.7.1986; Appl: 1.4.1987
  amending 83/229 EEC concerning the use of mono-ethylene glycol (MEG) and di-ethylene glycol (DEG) in regenerated cellulose film.

- 89/109 EEC (JOL 40 – 11.2.1989); Adopt: 21.12.1988; Appl: 10.7.1990
  on the approximation of the laws of the member states relating to materials and articles intended to come into contact with foodstuffs: new directive.

- 0/128 EEC (JOL 75 – 21.3.1990); Adopt: 23.2.1990; Appl: 1.1.1991
  relating to the monomers of plastic materials and articles intended to come into contact with foodstuffs.

2 The EEC packaging-waste legislation

The EEC policy is based upon the treaty of Rome (1957) modified by the Single European Act (1987). The following directives related to waste have been adopted:

- 75/442 EEC: framework directive on waste
- 78/319 EEC: directive on toxic and dangerous waste
- 84/631 EEC: directive concerning transfrontier shipment of hazardous waste
- 85/339 EEC: directive concerning containers of liquids for human consumption
- 86/279 EEC: amending 84/631
- 91/156 EEC: amending 75/442. The framework directives provide definitions of waste, waste prevention, recovery and disposal. They introduce the 'proximity'–principle and the 'polluter pays'–principle.

A first outline proposal for a directive on packaging was made available for discussion on 15 April 1991. The main objectives are:

- Prevent waste: within five years, limit the annual production of packaging waste per capita to the average per capita output of the EEC in 1990.

- Packaging waste recovery targets, to be reached within five years of directive implementation data (percentages by weight):

  - material recycling: 60% min. (60% by weight of each material should be recycled within 10 years, remove 90% by weight of packaging from waste stream for recovery);

  - energy recycling: 30% max;

  - landfilling: 10% max.

- Limit concentration of heavy metals and other hazardous substances in packaging materials.

- Encourage the use of returnable and recyclable packaging by means of economic and fiscal incentives.

- Inform consumers about environmental advantages of returnable and recyclable packagings by means of labelling systems.

Directive 85/339 EEC on containers of liquids for human consumption induced a variety of national decrees throughout Europe: European countries could be grouped into three categories according to the stringency of their legislation on waste management:

A: countries with stringent legislation on waste (Germany, Italy, Netherlands, Denmark)

B: countries with voluntary agreements and/or strong emerging legislation (France, Belgium, Sweden, United Kingdom, Austria, Switzerland)

C: other countries (Spain, Portugal, Ireland, Finland, Norway, Greece)

The current status of national legislation (May 1992) of European countries concerning waste management is summarised in the following overview:

		Areas									
	Countries	Bans	Obligations	Mandatory deposit	Recycling policy	Re-use	Taxes	Voluntary fee	Legal/Voluntary framework	Official targets	Hazardous substances
Denmark	DK	B	B	B	–	B	B	–	B/C	C	P
Germany	D	–	A	P	A	B	–	A	A	A	C
Italy	I	–	P	–	A	–	P	–	B	B	–
Netherlands	NL	–	C	C	A	P/C	C	–	A	A	C
Austria	A	C	C	B	–	–	–	–	A	B	C
Belgium	B	C	C	–	A	–	–	A	A	A	C
France	F	C	B/C	–	P	P	–	–	P	P	–
Sweden	S	C	B	B	B/C	C	B	–	P/C	C	C
Switzerland	CH	P/C	B	C	–	–	C	–	–	P	P
UK	GB	–	–	–	–	–	–	–	C	–	–
Greece	GR	–	–	–	P	–	–	–	C	–	–
Ireland	IRL	C	–	–	–	–	–	–	–	–	–
Norway	N	–	–	B	C	C	B	–	C	C	C
Portugal	P	–	–	–	–	–	–	–	–	C	–
Spain	E	–	–	–	–	–	–	–	–	–	–

A: all packaging included; P: partially included; B: only for beverages/liquids;
C: measures under consideration

Appendix 4

List of some EAN acknowledged prefixes

Prefix	Organization/Countries
00–09	UCC (USA and Canada)
20–29	Numberings in closed environments
30–37	GENCOD (France)
40–43	CCG (Germany)
49	Distribution Code Centre (Japan)
50	ANA Ltd (UK) and ANA of Ireland (Ireland)
520	HELLCAN (Greece)
54	ICODIF (Belgium and Luxembourg)
560	CODIPOR (Portugal)
57	Dansk Varekode Administration (Denmark)
590	Bar Coding Centre of Poland (Poland)
600–601	SAANA (South Africa)
64	Central Chamber of Commerce (Finland)
70	Norsk Varekodeforening (Norway)
73	Swedish EAN-Committee (Sweden)
76	SACV (Switzerland)
789	ABAC (Brazil)
80–83	INDICOD (Italy)
84	AECOC (Spain)
87	STICHTING UAC (The Netherlands)
90–91	EAN – Austria (Austria)
93	APNA Ltd (Australia)
94	NZPNA Ltd (New Zealand)
978–979	Books (ISBN)
98-99	Coupons, voucher

[Source : ICODIF – Belgium]

Appendix 5

LIST OF FOOD ADDITIVES

E-100 series: colourants
as full colouring:

E 102:	tartrazine (yellow)
E 120:	cochineal (red)
E 127:	erythrosine (red)
E 131:	patent blue V (blue)
E 140:	chlorophyll (green)

as surface colouring

E 170:	calcium carbonate
E 173:	aluminium

as special surface colourants

E 180:	pigment rubine for colouring surface of cheese

E-200 preservatives

E 200:	sorbic acid
E 201:	sodium sorbate
E 210:	benzoic acid
E 211:	sodium benzoate
E 214:	ethyl 4-hydroxybenzoate
E 221:	sodium sulphite
E 250:	sodium nitrite
E 251:	sodium nitrate

E-300 series: antioxidants

E 300:	L-ascorbic acid
E 301:	sodium l-ascorbate
E 320:	butylhydroxyanisol (BHA)

E 321: butylhydroxytoluene (BHT)
E 330: citric acid
E 331: sodium citrate

E-400 series: stabilizers, emulsifiers
E 401: sodium alginate
E 406: agar
E 414: arabic gum
E 440: pectin
E 466: sodium carboxymethylcellulose

E-500 series: flavourings
B 500: coumarin
B 540: caffeine
B 541: kinin

E-600 series: anticoagulants
B 600: silicon dioxide
B 644: calcium stearate

E-700 series: acids, bases and salts
B 720: maleic acid
B 730: gluconic acid
B 771: sodium carbonate

E-800 series: coatings
B 802: carnauba-wax
B 805: paraffin

E-900 series: gases
B 901: nitrogen
E 290: carbon dioxide

Appendix 6

PACKAGING RESEARCH INSTITUTES

Australia
Packaging Training and Research & Development Centre
PO Box 64, Footscray, Victoria, 3011 Australia
Tel: +61 (0)3 688 4244

Austria
Verpackungslabor für Lebensmittel und Getranke/Gregor
Mendelstrasse 33, A1180 Wien
Tel: +43 1 342500

Belgium
Belgian Packaging Institute
Picardstraat 15, B1210 Brussels
Tel: +32 (0)2 427 25 93

Brazil
Food Packaging Technology Centre
Av. Brazil 2880, CEP 13073 Campinas Sao Paulo
Brazil
Tel: +55 (0)192 415222

Czechoslovakia
IMADOS
Konevova 131, Zizkov, 13083 Praha 3 –
Tel: +42 (0) 2 264 1419

Denmark
The Danish Packaging and Transportation Research Institute (ETI)
Gregersenvej
PO Box 141 DK-2630 Taastrup
Tel: +45 42 9966 11

Finland
The Finnish Pulp and Paper Research Institute
PO Box 70, SF 02151 ESPOO, Helsinki 10
Tel: +358 90 43711

France
National Testing Laboratory (LNE)
Avenue Enrico Fermi, 5, F.78190 Trappes
Tel: +33 1 3069 1000

-ADRIAC- Reims
Esplanade Roland Garros, Pole Technologique Henri Farman
F-51100 Reims
Tel: +33 2684 5454

Germany
Packaging Development and Testing Centre (BFSV)
Lohbrugger Kirchstrasse 65, D2050 Hamburg 80
Tel: +49 (0)40 72522756

Institute for Food Technology and Packaging (ILV)
Schragenhofstrasse 35-D-8000 Munchen 50
Tel: +49 (0)89 149009-0

Federal Institute for Materials Research & Testing,
BAM unter den Eichen 87, D1000 Berlin 45
Tel: +49 (0)30 8104-1

Hungary
Institute of Materials Handling and Packaging (ACSI)
PO Box 189, H1431 Budapest
Tel: +36 (0)1 113 7460

Israel
Technion, Israel Institute of Technology
Technion City, Haifa 32000, Israel
Tel: +972 (0)4 292451

Italy
Research Centre for Industrial Technology (SIVA)
via Milano 1, I-20085 Locate triulzi, Milano
Tel: +39 (0)2 9077864

Japan
Japan Packaging Institute
7F.Kashikiehi Ningyo-cho Bldg
3-10-1 Ningyo-cho Nihonbashi Chuo-ku, Tokyo
Tel: +81 (0)3 3249 6301

Mexico
National Laboratories for Industrial Development (LANFI)
Av. Industria Militar 261, 53390 Naucalpan Edo de Mexico
Tel: +52 (0)5 5890199

Netherlands
TNO-Centre for Packaging Research
Schoenmakerstraat 97, PO Box 6034, Nl2600 J.A. Delft
Tel: +31 (0)15 696487

Norway
Norwegian Food Research Institute (MATFORSK)
Osloveien 1-N-1430 AAS
Tel: +47 (0)9 970 100

Poland
Polish Packaging Research and Development Centre (PRDC)
Ulica Konstancinska 11, P-02942 Warsrawa
Tel: +48 (0)22 422011

Sweden
Swedish Packaging Research Institute (PACKFORSK)
Torshamnsgatan 24, Box g, S-16493 Kista, Stockholm
Tel: +46 (0)8 7525700

Switzerland
Swiss Federal Laboratories for Materials Testing
and Research (EMPA)
Unterstrasse 11, CH9001 St. Gallen
Tel: +41 (0)71 209141

United Kingdom
Pira International
Randalls Road, Leatherhead, Surrey KT22 7RU
Tel: +44 (0)372 376161

United States of America
Michigan State University, School of Packaging
East Lansing, Michigan 48824-1223
Tel: +1 517 3536542

Rochester Institute of Technology,
Department of Packaging Science
Tel: +1 716 4755557

State University of New Jersey,
Centre for Packaging Science and Engineering
Bush Campus Building 3529, Piscataway, New Jersey 08855
Tel: +1 908 9323224

Bibliography

1 **Adcock, L H** *Migration from Plastics into Food,* Infopachs Training Course, 1981

2 **Adcock, L H, Hope, W G and Paine, F A,** 'The migration of non-volatile compounds from plastics – Part I' *Plastics and Rubber* Vol 5 No 1 (1980) pp 27–43

3 **Adcock, L H, Hope, W G and Paine, F A,** 'The migration of non-volatile compounds from plastics – Part II' *Plastics and Rubber* Vol 5 No 2 (1980) pp 71–77

4 **Ahlhaus, O** 'Problems and results in connection with the recycling of commodities and packaging materials', *in* **Lox, F (Ed.)** *ISCAT II* (1979)

5a **Ahlhaus, O, Boldt, G and Klein, K** *Taschenlexikon Umweltschuz* Pädagogischer Verlag Schwann, Düsseldorf (1979)

5b **Ahlhaus, O, Boldt, G, Gonsior, B, Klein, K L and Ziburske, H** *Taschenlexikon Energie* Pädagogischer Verlag Schwann, Düsseldorf (1981)

6 **Arvidsson, L, Edholm, B, Hocking, C, Jöhnson, G and Lidgren, K** *Energy Consumption in the Packaging Sector* Swedish Packaging Research Institute, Spånga (1978)

7 **Ashworth, B T and Hill, P** 'Some observations on unusual antioxidant effects' *Kautschuk und Gummi-Kunststoffe* Vol 34 No 1 (1981) pp 18–22

8 **Autian, J** 'Plastics in parenteral packaging' *The Bulletin of the Parenteral Drug Association* Vol 14 No 4 (1960) pp 10–27

9 **Autian, J** 'Plastics in pharmaceutical practice and related fields – Part I' *Journal of Pharmaceutical Sciences* Vol 52 No 1 (1963) pp 1–23

10 **Autian, J** 'Plastics in pharmaceutical practice and related fields – Part II' *Journal of Pharmaceutical Sciences* Vol 52 No 2 (1963) pp 105–121

11 **Autian, J** *The Use of a Biological Screen for the Detection of Toxic Leachables in Plastics* University of Tennessee, Memphis

12 **Autian, J and Kapadia, A J** 'A note on the leaching of a constituent from medical grade plastic tubings' *Drug Standards* Vol 28 No 4 (1960) pp 101–103

13 **Autian, J Ph.D** 'Plastic packaging and cosmetic products' *Proc. of the Scientific Section of the Toilet Goods Association* Vol 43 (1965) pp 14–20

14 **Bailey, H C** 'Chemistry of antioxidants, antiozonants and heat stabilizers' *in* **Pinner, S H (Ed.)** *Weathering and Degradation of Plastics* Columbine Press, Manchester & London (1966) pp 35–49

15 **Bakker, P J** 'Kunststoffen en het milieucomplex' *Plastica* Vol 29 No 12 (1976) pp 409–421

16 **Bengough, W I** 'Thermal degradation of polyvinyl-chloride' *in* **Pinner, S H (Ed.)** *Weathering and Degradation of Plastics* Columbine Press, Manchester & London (1966) pp 66–81

17 **Berg, H F, Guess, W L and Autian, J** 'Interaction of a group of low molecular weight organic acids with insoluble polyamides I' *Journal of Pharmaceutical Sciences* Vol 54 No 1 (1965) pp 79–84

18 **Bobelijn, L** *Verpakking van Levensmiddelen* thesis HRIBP Gent (1979)

19 **Bontinck, W J** 'Voedingshygiënische aspecten van verpakkingsmaterialen' *Het Ingenieursblad* Vol 47 No 10 (1978)

20 **Boonet, E** 'The European Article Numbering on the packings' *in* **Lox, F (Ed.)** *ISCAT II* (1979) A3.1–A3.6

21 **Botrel, J** 'Problémes posés par le controle de la migration des matériaux plastiques au contact des aliments' *Annales des Fabrications, de l'Expertise Chimique et Toxicologique* Vol 75 No 805 (1982) pp 75–80

22 **Boustead, I and Hancock, G F** *Energy and packaging* Ellis Horwood Ltd, Publishers, Chichester, England, (1981)

23 **Briston, J H and Katan, L L** *Plastics in Contact with Food* Food Trade Press Ltd, London (1974)

24 **Buekens, A** 'Technische methoden van milieubescherming' *Intermediair* Vol 5 No 15 (1974) 3, pp 17–19

25 **Buekens, A** *Vaste Afvalstoffen, ontstaan, Verwerking en Beheer* Monografieen Leefmilieu Nu, De Nederlandsche Boekhandel, Antwerpen (1975)

26 **Bussink, J** 'Ontwikkelingen in polymere materialen' *Plastica* Vol 34 No 6 (1981) pp 158–162

27 **Calzolari, C C** 'Der mensch und die energiefrage' *Forum Ware* Vol 9 (1981) No 1–2 pp 17–25

28 **Caputo, D L** 'Municipal plastic waste: a largely untapped resource' *Plastics Engineering* (February 1982) pp 33–35

29 **Carswell, T S and Nason, H K** *Modern Plastics* 21 (1944) p 121

30 **Chesney, J C M** *Packaging of Cosmetics and Toiletries* Newnes-Butterworths, London (1974)

31 **Cichon, M and Wlodarczyk, W** *Towaroznawstwo Opakowan* Wyrsza Szkola Ekonomiczna W Krakowie, Krakow (1972)

32 *Pharmaceuticals* WHO, Genéve (1974)

33 **Crosby, N T** *Food Packaging Materials* Applied Science Publishers, London (1981)

34 **Daun, H and Gilbert, S G** 'Migration of plasticizers from polyvinylchloride packaging films to meat' *Journal of Food Science* Vol 42 No 4 (1977) pp 561–562

35 **De Leenheer, A and Heyndrickx, A** 'Onderzoek op de toxicologische aanvaarding van organische polymeren voor waterdistributiesystemen voeding en farmaceutische verpakkingen' *Pharmaceutisch Tijdschrift voor Belgiä* Vol 43 No 4 (1966)

36 **De Maeseneer, J** *Biologische Aspekten van het Waterzuiveringsproces* Kursus-Interfacultair Centrum voor de studie van lucht-, bodem- en waterverontreiniging, Rijksuniversiteit te Gent (1975)

37 **De Marco, O and Mazzacane, E M P** 'The knowledge of commodities in Caius Plinius's natural history on the 1900th anniversary of the author's death' *Forum Ware* Vol 7 No 3–4 (1979) pp 187–190

38 **De, S K and Autian, J** 'Permeation of phenol and substituted phenols in aqueous solutions through polyurethane film' *Indian Journal of Chemistry* Vol 13 (April 1975) pp 366–368

39 **De Smet, R** *Voorverpakte Geportioneerde Vleeswaren* thesis HRIPB Gent (1982)

40 **De Vos, P** 'Het afvalverwerkingshedrÿff van de stad Gent' *Leefmilieu* Vol 5 No 1 (1982) pp 3–12

41 **De Vos, P** 'Het composteren van huisvuil' *Beswa Revue* Vol IX No 1 (1978)

42 **De Wilde, J H** *Analytical Problems Created by Packaging Material* Toxicology Division Shell International Research My. NV, The Hague (1966)

43 **De Winter, A** *Plastic-verpakking van Levensmiddelen* thesis HRIPB Gent (1978)

44 **De Witte, A** *Verpakking van Levensmiddelen en Haar Economische Betekenis* thesis HRIHAN, Aalst (1981)

45 **Dikkers, G J O D** 'Kostenfaktoren in de kunststofverwerking' *Plastica* Vol 31 No 8 (1978) pp 250–253

46 **Downes, T W and Gilbert, S G** 'The effect of surface treatments on the migration of organotin stabilizer from polyvinylchloride films to contacting liquid media' *Deutsche Lebensmittel – Rundschau* Vol 71 No 12 (1975) pp 422–425

47 **Dumeez, M** 'Resources and collection of waste paper' *in* **Lox, F (Ed.)** *ISCAT II* (1979) C5.1–C5.16

48 **Elskens, I** *Impacten op Aquatische Biotopen van de Uitbreiding van Kerncentrales* Lezingen over menselijke ecologie, VUB, Brussels (1981)

49 **Enstra, W C** 'Peroxid – vernetzung von polymeren' *Kautschuk und Gummi-Kunststoffe* Vol 32 No 10 (1979) pp 756–759

50 **Figge, K** 'Migration of additives from plastic films into edible oils and fat simulants' *Fd. Cosmet. Toxicol.* Vol 10 (1972) pp 815–828

51 **Figge, K and Koch, J** 'Effect of some variables on the migration of additives from plastics into edible fats' *Fd. Cosmet. Toxicol.* Vol 11 (1973) pp 975–988

52 **Fink, P** 'Evaluation of packaging systems in relation to the environmental impact' *in* **Lox, F (Ed.)** *ISCAT II* (1979) C6.1–C6.16

53 **Fink, P** *Technologische Aspecten der Distribution*, Cursus Hochschule St Gallen, Switzerland

54 *The Fläkt RRR-Plant at Lövsta* Mededeling van het 'Stockholms kommun Miljö – och Hälsovårds – Förvaltingen' (1978)

55 **Gabbay, Y and Zilkha, A** 'Synthesis of potential biologically active polymers based on polystyrene' *Israel Journal of Chemistry* Vol 17 (1978) pp 304–306

56 **Garlanda, T and Masoero, M** 'Considerazioni sulla migrazione di componenti di materie plastiche in solventi a contatto' *La Chimica e l'Industria* Vol 48 No 9 (1966) pp 936–943

57 **Ghatge, N D and Vaidya, S V** 'Organotin Compounds in Stabilization of PVC' *Kautschuk und Gummi-Kunststoffe* Vol 32 No 4 (1979) pp 254–257

58 **Giacin, J R** 'Evaluation of plastic packaging materials for food packaging application' *Journal of Food Safety* No 4 (1980) pp 257–276

59 **Giacin, J R** 'Tracking down those toxics with gas chromatography and mass spectroscopy' *Package Engineering* (1980) pp 70–79

60 **Gilbert, S G** 'A scientific basis for regulation of indirect food additives from packaging materials' *Journal of Food Quality* No 2 (1979) pp 251–256

61 **Gilbert, S G** 'Migration – entropic considerations' *in* **Lox, F (Ed.)** *ISCAT II* (1979) B1.1–B1.11

62 **Gilbert, S G** 'Modelling the migration of indirect additives to food' *Food Technology* (April 1979) pp 63, 65
63 **Gilbert, S G, Miltz, J and Giacin, J R** 'Transport considerations of potential migrants from food packaging materials' *Journal of Food Processing and Preservation* No 4 (1980) pp 27–49
64 **Giorgio, N** 'Warenkunde/merceologia: The unifying ground for economic and natural sciences' *Forum Ware* Vol 9 No 1–2 (1981) pp 28–30
65 **Goff, J W and Twede, D** *Boxes Bags and Cans* Michigan State University, School of Packaging, Special Report No 14 (1979)
66 **Goldan, D** 'Glass containers and the energy gap' *Forum Ware* Vol 4 No 1 (1976)
67 **Goldan, D** 'Reduction of commodity spoiling and saving resources by sophisticated packaging' *Forum Ware* Vol 6 No 1–2 (1978) pp 101–104
68 **Gonzales, M A, Nematollahi, J, Guess, W L and Autian, J** 'Diffusion, permeation and solubility of selected agents in and through polyethylene' *Journal of Pharmaceutical Sciences* Vol 56 No 10 (1967) pp 1288–1293
69 **Grandry, G** 'Prospection du marché du compost dans le cadre du recyclage des déchets organiques' *Rapport SDRW* (1978)
70 **Gross, D** 'Elastomere im trinkwasserkontakt' *Kautschuk und Gummi-Kunststoffe* Vol 33 No 12 (1980) pp 997–999
71 **Grundke, G** 'Atmospheric influences on commodities and packages and their significance for the international trade' *in* **Lox, F (Ed.)** *ISCAT II* (1979) C1.1–C.1.9
72 **Grundke, G** *Grundriss der allgemeinen Warenkunde* Band III VEB Fachbuchverlag Leipzig (1981)
73 **Grundke, G** Über Erfahrungen mit Lebensmittelverpackungen in den feuchten Tropen, Ernährungsforschung, Band XI, 3 (1964)
74 **Haesen, G and Schwarze, A** *Migration Phenomena in Food Packaging* Commission of the European Communities, Brussels/Luxembourg 1978

75 **Haesen, G, Depaus, R, van Tilbeurgh, H, Le Goff, B and Lox, F** 'The effect of γ-irradiation on the migration behaviour of Advastab 17 MOK in PVC' *J. Ind Irrad. Tech.* Vol 3 No 1 (1983) pp 259–260

76 **Hannon, B M** 'Bottles cans energy' *in* **Holloway, R J and Hancock, R S (Eds.)** *The Environment of Marketing Management* John Wiley and Sons, New York (1974) pp 90–100

77 **Heiss, R** *Verpackung von Lebensmitteln* Springer-Verlag, Berlin (1980)

78 **Helen, H** 'The effect of packaging material on the shelf-life of biscuits. A critical problem, description and imperial investigations' *Forum Ware* Vol 6 No 1–2 (1978)

79 **Hens, L J J and Susanne, C** 'Ioniserende straling en het menselijk genoom' *Natuur en Techniek* Vol 45 No 7 (1977) pp 496–511

80 **Hobson, P N and Robertson, A M** *Waste Treatment in Agriculture* Applied Science Publishers, London (1977)

81 **Holloway, R J and Hancock, R S (Eds.)** *The Environment of Marketing Management* John Wiley and Sons, New York (1974)

82 **Hung, G W C and Autian, J** 'Use of thermal gravimetric analyses in sorption' *Journal of Pharmaceutical Sciences* Vol 61 No 7 (1972) pp 1094–1098

83 **Huyghebaert, A** 'Migration of styrene-monomer into food' *in* **Lox, F (Ed.)** *ISCAT II* (1979) B7.1–B7.8

84 **Iwaki, R** 'Commodity science' *Seirinshoin – Shinsha* 5 March 1976 p 245 (Japanese)

85 **Jönson, G** 'Food distribution in Europe' *Proc. of the Seminar on International Developments in Food and Pharmaceutical Packaging* Michigan State University, School of Packaging, 4–5 May 1982

86 **Kaltenbac, H C** *Die Rolle von Produkt und Verpackung in der Marktkommunikation* Verlag W Girardet, Essen (1975)

87 **Kapadia, A J, Guess, W L and Autian, J** 'Interaction of weak organic acids with insoluble polyamides II' *Journal of Pharmaceutical Sciences* Vol 53 No 7 (1964) pp 720–725

88 **Kapadia, A J, Guess, W L and Autian, J** 'Interaction of weak organic acids with insoluble polyamides I' *Journal of Pharmaceutical Sciences* Vol 53 No 1 (1964), pp 28–34
89 **Karel, M** *Nutritional Evaluation of Food Processing* 2nd edn AVI Publishing Co (1975) pp 419–455
90 **Katan, L L** *Hazard Analysis* Infopachs Training Course (1981)
91 **Kersten, H** 'Polyvinylideenfluoride (PVDF) – in theorie en praktijk' *Plastica* Vol 30 No 3 (1977) pp 64–67
92 **Kimmerle, G and Prager, F H** 'Toxizitäat der verschwelungs- und verbrennungsprodukte verschiedener Chemiewerkstoffe und -fasern' *Kautschuk und Gummi-Kunststoffe* Vol 33 No 5 (1980) pp 354–359
93 **Klein, M** 'Energy – the challenge to resource science' *Forum Ware* 7, 1–2 (1979), pp 109–112
94 **Knibbe, D E** 'Theory of extraction of additives from plastics by swelling solvents' *Plastica* Vol 24 No 8 (1971) pp 358–363
95 **Kotler, P** *Marketing Management* Prentice-Hall (1967) p 282
96 **Labuza, T P** 'Moisture gain and loss in packaged foods' *Food Technology* (April 1982) pp 92–97
97 **Labuza, T P, Tannenbaum, S R and Karel, M** 'Water content and stability of low-moisture, and intermediate-moisture foods' *Food Technology* Vol 24 No 5 (1969) p 35
98 **Lamberu, Y and Nauwelaerts, F** 'Is het aanwenden van PVC als verpakkingsmateriaal schadelijk voor de gezondheid?' *Chemie Magazine* Vol 4 No 2 (1978) pp 5–7
99 **Langrak, I** 'Irradiation and packaging in relation to food preservation' *International Packaging Congress* Paris 19 November 1980
100 **Lawrence, W H** 'Phthalate esters: the question of safety' *Clinical Toxicology* Vol 13 No 1 (1978) pp 89–139
101 **Lawrence, W H and Tuell, S F** 'Phthalate esters: the question of safety' *Clinical Toxicology* Vol 15 No 4 (1979) pp 447–466

102 **Lestang, P** 'Harmonizing international commodity classifications' *in* **Lox, F (Ed.)** *ISCAT II* (1979) A7.1–A7.4 *Forum Ware* Vol 7 No 3–4 (1979) pp 151–152

103 **Lidgren, K, Jönson, G, Arvidsson, L, Backman, M and Hocking, C** *Energy and Packaging* Swedish Packaging Research Institute, Spånga (1981)

104 **Liptak, B G** *Environmental Engineers' Handbook, Volume 3: Landpollution* Chilton Book Company, Padnor (1974)

105 **Lox, F** 'Bepaling van de maximale totale migratie van additieven bij PVC-krimpfoliân' *Chemie Magazine* Vol 7 No 5 (1981) pp 14–18

106 **Lox, F** *Industry and Ecology* Cursus VUB (1982)

107 **Lox, F** 'Migration – isotherms on PVC-shrink foils used for food packaging' *Proc. of the Third International Symposium on Commodity Science and Technology (ISCAT II)* Krakow, 20–23 1981 pp 264–267

108 **Lox, F** 'Migration of components from plastic packaging materials into food simulants' *Proc. of the Seminar on International Developments in Food and Pharmaceutical Packaging* Michigan State University, School of Packaging, 4–5 May 1982

109 **Lox, F** 'Overall-migration on PVC-packaging foils in contact with food and simulant-mixtures' *in* **Lox, F (Ed.)** *ISCAT II* (1979) B13.1–B13.9

110 **Lox, F** 'Packaging for health in trade – topics in commodity science' *Proc. of the Second International Symposium on Commodity Science and Technology (ISCAT II)* BIWT Gent 19–21 (1979)

111 **Lox, F** 'Sociale, economische en energetische aspecten in verband met huisafval en zijn verwerking, *Economisch en Sociaal Tijdschrift* Vol 29 No 2 (1975) pp 167–178

112 **Lox, F** 'The iron-, vitamin C-, sodium- and nitrate-content of commercial babyfoods' *Forum Ware* Vol 9 No 3–4 (1981) pp 95–96

113 **Lox, F Haesen, G and Autian, J** 'A test concerning the toxicity of the total extract of γ-irradiated rigid PVC films' unpublished results

114 Martin, W J and Weiland, H *Proc. First International Symposium on Materials and Energy from Refuse*

115 Massart, D L, Deelstra, H and Hoogewijs, G *Vreemde Stoffen in Onze Woeding* De Nederlandsche Boekhandel, Antwerpen/Amsterdam (1980)

116 Meadows, D *Rapport van de Club van Rome. De granzen aan de Groei* Het Spectrum, Utrecht/Antwerpen (1972)

117 Mercier, J P 'Quelques aspects de la physique des hauts polyméeres' *Industrie Chimique Belge* Vol 30 No 8 (1965) pp 813–819

118 Meulenhoff, J S 'Di-(2-ethylhexyl)-phtalaat in menselijk plasma' *Pharmaceutisch Weekblad* Vol 109 (1974) pp 719–720

119 Mickwitz, G *Marketing and Competition* Helsingfors Finland, Centraltry Cheriet (1950) p 88

120 Mozisek, M and Bityska, V 'Diffusion einiger Weichmacher in Naturkautschuk' *Kautschuk und Gummi-Kunststoffe* Vol 34 No 6 (1981) pp 437–478

121 Murgescu, C and Radulescu, O 'Basic raw materials and the development of world economy' *Forum Ware* Vol 1 No 1–2 (1979) pp 113–120

122 Nalepa, W and Cichon, Z 'Researches concerning the quality evaluation of tinplate cans used for fruit and vegetable preserves' *in* **Lox, F** ***(Ed.)*** *ISCAT II* (1979) B12.1–B12.5

123 Nasim, K, Meyer, M C and Autian, J 'Permeation of aromatic organic compounds from aqueous solutions through polyethylene' *Journal of Pharmaceutical Sciences* Vol 61 No 11 (1972) pp 1775–1780

124 Nijssen, Th 'Toelaatbaar styreengehalte in de lucht in diverse landen in Europa, tijdens de produktie van GVK' *Plastica* Vol 33 No 2 (1980) pp 40–42

125 Oswin, C R *Plastic Films and Packaging* Applied Science Publishers, London (1975)

126 Paine, F *Fundamentals of Packaging* Blackie & Son, London (1962)

127 **Pairon, G and Sarens, J** *Etudes des Avantages et Désavantages des Systémes 'Retournables' et 'Non Retournables'* GB-INNO-BM, Brussels (1981)

128a **Pairon, G and Sarens, J** *Etude des Emballages Consignés et Non-Consignés. Secteur Envisage: Eaux et Limonades* GB-INNO-BM, Brussels (1980)

128b **Pairon, G and Sarens, J** *Etude des Emballages Consignés et Non-Consignés. Secteur Envisage: La Biere* GB-INNO-BM, Brussels (1980)

129 **Pairon, G and Sarens, J** *Etude des Emballages Consignés et Non-Consignés. Secteur Envisage: Lait* GB-INNO-BM, Brussels (1980)

130 **Patfoort, G A** *Polymers* Scientific Publishers, Gent, (1974)

131 **Pavoni, J L, Heer, J E and Hagerty, D J** *Handbook of Solid Waste Disposal* Van Nostrand Reinhold Co, New York (1975)

132 **Peereboom, J W C** 'Schadelijke effecten van weekmakers' *Chemisch Weekblad* Vol 70 No 13 (1974) pp 5–7

133 **Pezarro, L** 'Kunststoffen en weekmakers bij bloedtransfusie' *Plastica* Vol 33 No 2 (1980) pp 27–31

134 **Pinner, S H** *Modern Packaging Films* Butterworths, London (1967)

135 **Pinner, S H** *Weathering and Degradation of Plastics* Columbine Press, Manchester & London (1966)

136 **Poels, C L M** 'Toxicologische aspecten en beoordeling van glasvezelversterkie kunststofbuizen voor transport van drinkwater' *Plastica* Vol 33 No 12 (1980) pp 380–385

137 **Polack, A E, Nunez, L J and Autian, J** 'Transport of solutes into polyethylene bottles from aqueous solutions' *International Journal of Pharmaceutics* Vol 3 (1979) pp 157–175

138 **Purwadaria, H K, Heldman, D R and Kirk, J R** 'Computer simulation of vitamin degradation in a dry model food system during storage' *Journal of Food Process Engineering* Vol 3 (1979) pp 7–28

139 **Rasch, R** 'Emissionen aus Müllverbrennungsanlagen bei Müll- und Kunststoffverbrennung' *Müll und Abfall* Vol 3 (1972) pp 88–100
140 **Roberts, M S, Polack, A E, Martin, G and Blackburn, H D** 'The storage of selected substances in aqueous solution in polyethylene containers' *International Journal of Pharmaceutics* Vol 2 (1979) pp 295–306
141 **Roberts, M S, Cossum, P A, Galbraith, A J and Boyd, G W** 'The availability of nitroglycerin from parenteral solutions' *Journal of Pharmaceutical Pharmacol.* Vol 32 (1980) pp 237–244
142 **Rodell, M B, Bodnar, R, Guess, W L and Autian, J** 'Further studies on the interaction of sorbic acid with an insoluble polyamide' *Journal of Pharmaceutical Sciences* Vol 54 No 1 (1965) pp 129–134
143 **Rodell, M B, Guess, W L and Autian, J** 'Interaction of a group of weak organic acids and phenols with a polyamide' *Journal of Pharmaceutical Sciences* Vol 55 No 12 (1966) pp 1429–1434
144 **Rodell, M B, Guess, W L and Autian, J** 'Interaction of sorbic acid with an insoluble polyamide' *Journal of Pharmaceutical Sciences* Vol 53 No 8 (1964) pp 873–877
145 **Ross, C F** *Packaging of Pharmaceuticals* Newnes-Butterworths, London (1975)
146 **Sacharow, S and Griffin, R** *Food Packaging* AVI Publishing Co, Westport Connecticut (1970)
147 **Schliekelmann, R S** 'Een blik in de ontwikkeling van konstruktiematerialen in het verieden en in de toekomst II' *Plastica* Vol 34 No 11 (1981) pp 314–317
148 **Schönborn, H H** 'Arten, mengen und Zusammensetzung von Kunststoffen' *Müll und Abfall* Vol 1 (1972) pp 1–10
149 **Schuck, H** 'Die Gaspermeabilität von Hochpolymeren, insbesondere von Kautschuken' *Kautschuk und Gummi-Kunststoffe* Vol 33 No 9 (1980) pp 705–715
150 **Schuur, G** 'De milieuhygiänische aspecten van verpakkingsmaterialen en verpakkingen' *Plastica* Vol 25 No 1 (1972) pp 2–9

151 **Serota, D G, Meyer, M C and Autian, J** 'Effects of structure on permeability of substituted analines from aqueous solutions through polyethylene' *Journal of Pharmaceutical Sciences* Vol 61 No 3 (1972) pp 416–419

152 **Singh, R P** 'Kinetics of quality degradation: ascorbic acid oxidation in infant formula during storage' *Journal of Food Science* Vol 41 (1976) pp 304–308

153 **Singh, R P, Heldman, D R** 'Simulation of liquid food quality during storage' *ASAE* Vol 19 No 1 (1976) pp 178–180, 182–184

154 **Singh, R P, Heldman, D R and Kirk, J R** 'Kinetic analysis of light-induced riboflavin loss in whole milk' *Journal of Food Science* Vol 40 (1975) pp 164–167

155 **Smeyers, E** *Mutageniciteit, Cancerogeniciteit en Teratogeniciteit van Conserveermiddelen* Proefschrift VUB, (1981)

156 **Souffie, R D** 'Segmented polyether-ester copolymers are thermoplastic elastomers' *Plastics Engineering* (March 1982) pp 21–24

157 **Squires, S** 'Injection moulding – quo vadis? (II)' *Plastica* Vol 34 No 3 (1981) pp 64–66

158 **Staudinger, J J P** *Disposal of Plastics Waste and Litter* Society of Chemical Industry, London (1970)

159 **Staudinger, J J P** *Plastics and the Environment* Hutchinson, London (1974)

160 **Stevens, M P** *Polymer Chemistry* Addison-Wesley Publishing Co, Massachusetts (1975)

161 **Strahler, A N and Strahler, A H** *Environmental Geoscience: Interaction between Natural Systems and Man* Hamilton Publishing Company, Santa Barbara (1973)

162 **Sundström, G** *Melkverpakkingen en Energie* Malmo, Sweden (1980)

163 **Swinbank, C** *Packaging of Chemicals* Newnes-Butterworths, London (1973)

164 **Tetart, P, Haguenoer, J M, Erb, F and Dequidt, J** 'Migration of monomers and metals from plastics to milk and its derivatives' *in* **Lox, F (Ed.)** *ISCAT II* (1979) B11.1–B11.21

165 **Vanacker, L** *Marktprospectie van Compost in het Kader van de Recyclage van Organische Afvalstoffen* Rapport R & D, Gent (1978)

166 **Van Cleemput, O and Baert, L** 'Voedselkonservering door bestraling' *Het Ingenieursblad* Vol 51 No 8 (1982) pp 190–195

167 **Van den Broeck, M** *De Selektieve Ophaling van oud Papier en Glas in Belgiä* Proefschrift VUB (1978)

168 **Van der Kuil, J A** 'Verpakking als afvalprobleem' *Chemisch Weekblad* Vol 68 No 10 (1972) pp 15–16

169 **Van der Linden, R and Steffen, K C** 'Migration through a matrix, practical realisations and discussions for polyethylene plastics' *in* **Lox, F (Ed.)** *ISCAT II* (1979) A4. 1–A4.21

170 **Vanderpoorten, R** 'Kunststofverpakkingen en de migratie in zuivelprodukten' *Landbouwtijdschrift* Vol 27 No 6 (1974) pp 1491–1510

171 **Van Vucht, H A** 'Achtergronden van de huidige EEG-regelingen voor vinylchloride' *Plastica* Vol 33 No 3 (1980) pp 70–74

172 *Verpakking Recycling* Belgisch Verpakkingsinsituut, Brussels (1978)

173 **Verstraete, W** 'Biodegradeerbare plastiekverpakkingen, een wensdroom' *Intermediair* Vol 5 No 7 (1974) pp 3–4

174 **Verstraete, W** *Verwerken van Huishoudelijke en Landbouwtechnische Afvalstoffen* RUG, Interfacultair Centrum Milieusanering (1974)

175 **Viatour, G** 'Het hercycleren van plastic afvalstoffen in Belgiä' *Behandeling Verpakking* (March 1976) pp 34–42

176 **Vink, P** 'Sterilisatie van kunststoffen en rubbers' *Plastica* Vol 34 No 5 (1981) pp 123–127

177 **Vogel, G** *Der Beitrag des Recyclings zur Stabilisierung des Technosozio-öko-nomischen Systems* Institut für Technologie und Warenwirtschaftslehre der Wirtschaftsuniversität, Wien (1978)

178 **Vogel, G** *Sammlung von Alstoffen* Institut für Technologie und Warenwirtschaftslehre der Wirtschaftsuniversität, Wien (1980)

179 **Vogel, G** *Simulationen von Altstoffsammlungen* Institut für Technologie und Warenwirtschaftslehre der Wirtschaftsuniversität, Wien (1980)

180 **Vom Bruck, C G, Figge, K and Rudolph, F** 'Interaction of fat containing food with plastic packaging' *Voordracht ISF-AOCS-Congress* New York (1980)

181 **Wallhäusser, K H** 'Die Bezeitigung von Kunststoffe in der geordneten Deponie' *Müll und Abfall* Vol 1 (1972) pp 10–21

182 **Wanninger, L A** 'Mathematical model predicts stability of ascorbic acid in food products' *Food Technology* (June 1972) pp 42–45

183 **Willems, N** *Verbrandingstechnische Aspekten bij Verbranding van Huisvuil* Proefschrift VUB (1981)

184 **Winkler, P** *Encyclopedie der Materialenkennis Samengesteld onder Leiding* **Berg, K, Boes, G and van der Linde, J (Eds.)** Agon Elsevier Amsterdam/Brussel (1942)

185 *Witboek, Verpakking-Economie-Ecologie* Belgisch Verpakkingsinstituut, Brussel (1977)

186 **Zielhuis, R L** 'Nieuwe inzichten in de gezondheidsrisico's bij blootstelling aan styreen' *Plastica* Vol 32 No 12 (1979) pp 396–401

187 APME 'Information system on plastics recycling in Western Europe' SEMA-Group (1991)

188 **Bakker, M** *The Wiley Encyclopaedia of Packaging Technology* John Wiley and Sons, New York (1986)

189 **Bojkow, E** *Getränke – Verpackung und Umwelt* Springer-Verlag Wien, New York (1989)

190 **Bureau, G and Multon, J L** *L'emballage des Denrées Alimentaires de Grande Consommation* Lavoisier, Paris (1989)

191 **Crosby, N T** *Food Packaging Materials* Applied Science Publisher Ltd (1981)

192 **Danger, E P** *Selecting Colour for Packaging* Gower Technical Press (1987)

193 **Deboyser, K** *Kunststofverpakking; een Milieuprobleem?* Thesis VUB (1990)

194 **De Caevel, B** *Life-Cycle Analysis van PVC* Thesis VUB (1991)

195 **Delmonte, J** *Origins of Materials and Processes* Technomic Publishing Co, USA (1985)

196 **EEC** – Proposal for a council regulation concerning Eco-label, Draft document of the Working Group, Rome (July 1991)

197 **Feigenbaum, A and Rossi, L** 'La reglementation Europêenne des emballages alimentaires: les materiaux plastiques' *Ann. Fals. Exp. Chim.* Vol 84, No 894 (1991) pp 21–25

198 **Feigenbaum, A, Ducruet, V J, Delpal, S and Wolff, N** 'Food and packaging interactions: penetration of fatty food simulants into rigid PVC' *Journal of Agricultural and Food Chemistry,* Vol 39 (1991) pp 1927–32

199 **Habesatter, K and Widmer, F** *Ockobilanz von Packstoffen, Stand 1990* Buwal Bern (1991)

200 **Hocking, M B** Relative merits of PS-foam and paper in hot drink cups: implications for packaging *Environmental Management,* Vol 15, No 6 pp 731– 747 (1991)

201 **Hotchkiss, J H** 'Food and packaging interactions', *ACS-Symposium Series 365* Washington (1988)

202 **Ferranti, M P and Ferro, G L** *Sorting of Household Waste and Thermal Treatment* Elsevier Applied Science Publishers (1985)

203 **Fink, P** *Der Beitrag des Naturwissenschaftlers zur Bewältigung der Umweltprobleme* Wirtschaftsuniversität Wien (1991) The contribution of science to the solution of environmental problems

204 **Grundke, G** *Warenschäden im Modell* VEB Fachbuchverlag Leipzig (1982)

205 **Hauschild, G and Spingler, E** *Migration bei Kunststoffverpackungen* APV, Stuttgart (1988)

206 **Jackson, F R** *Recycling and Reclaiming of Municipal Solid Wastes* Noyes Data Corporation (1975)

207 **Kadoya, T** *Food Packaging* Academic Press Inc, London (1990)

208 **Klompmaker, M** 'Convenant verpakkingen getekend' *Pack. News* Vol 4 (1991) pp 41–64

209 **Kondyli, E, Demertzis, P G and Kontominas, M G** 'Migration of di-octylphtalate and di-octyladipate plasticizers from PVC-films into olive-oil' *Food Chemistry* Vol 36 (1990) pp 1–10

210 **Kooijman, J M** 'Ketenbeoordeling van verpakkingen, milieuaspekten' *Voedingsmiddelen Technologie* Vol 1 (1992) pp 31–34

211 **Kroschwitz, J I** *Concise Encyclopedia of Polymer Science and Engineering* John Wiley & Sons, New York (1990)

212 *Life-cycle Analysis: Proc. of the Workshop* B-Leuven, 24-25 Sept, Procter & Gamble (1990)

213 **Lox, F** *Verantwoorde Verpakking* Monografie No 22, Nederlandsche Boekhandel, Antwerpen (1983)

214 **Lox, F and Walden, A** *Proc. of the 16th IAPRI-Symp.* Gent, 1–4 May 1988, BIWT-IBTM, Gent (1989)

215 **Lox, F and Naeyaert, A** 'Hoe veilig zÿn kinderveilige verpakkingen' *in* **Verhellen, E (Ed.)** *Rechten van Kinderen* (1989) pp 349–369 Kluwer rechtswetenschappen, Antwerpen

216 **Maes, M** *La Maîtrise des Déchets Industriels* P Joharet and Fils, Paris (1991)

217 **Mercer, A, Castle, L, Comyn, J and Gilbert, J** 'Evaluation of a predictive mathematical model of di-(2 ethylhexyl) adipate plasticiser migration from PVC-film into foods' *Food Additives and Contaminants* Vol 7 No 4 (1990) pp 497–507

218 **Mewis, J** 'Gevaarlÿke stoffen', Monografie No 21 Stichting Leefmilieu, De Nederlandsche Boekhandel, Antwerpen (1983)

219 **Moran, Morgan, Wiersma** *Introduction to Environmental Science* Freeman Company, New York (1986)

220 **Paine, F** *The Packaging User's Handbook* Blackie & Son, UK (1991)

221 **Paine, F and Paine, H** *A Handbook of Food Packaging* Leonard Hill, Blackie & Son, UK (1992)

222 **Paine, F** *Packing Design and Performance* Pira, Leatherhead, UK (1990)
223 **Paine, F** *Modern Processing, Packaging and Distribution Systems for Food* Blackie & Son, UK (1987)
224 **Pascat, B and Bureau, G** 'Food packaging; a burden or an achievement' *Symposium* Reims 5-7 June 1991, ADRIAC-REIMS (Oct 1991)
225 **Piringer, O** 'Wechselwirkungen zwischen Lebensmittel und Verpackung' *Chem. Imp. Techn.* Vol 60, No 4, (1988) pp 255–265
226 **Ricklefs, R** *Ecology* Freeman Company (1990)
227 **Robinson, W** *The Solid Waste Handbook* John Wiley & Sons, UK (1986)
228 **Roge, E and Roge, T** 'Het kurken; l'Ingéniosité du bouchage' *Pack. News,* Vol 4 (1991) pp 20–35
229 **Rijsdorp, I, Guinee, J and Huppes, G** *Milieu-effekten van Huishoudelÿke Verpakkingen* Deel I and II, Centrum voor milieukunde-Leiden (1989)
230 **Rombouts, C** *Afvalverwerking en Recyclage van Kunststofverpakkingen* Thesis, UIA (1989)
231 **Seitz, K** *Polymers and the Environment* Exxon Chemical (1991)
232 **Sutcliffe, C** *The Dangers of Low Level Radiation* Blackmore Press (1987)
233 **Taverdet, J, Vergnaud, J** 'Modelization of matter transfers between plasticised PVC and liquids in case of a maximum for liquid-time curves' *Journal of Applied Polymer Science* Vol 31 (1986) pp 111–122
234 **Vanthourout, A** *Packaging Handbook* Belgian Packaging Institute (1984)
235 **Veaux, M** *Eco-Labelling, Technical Notice* LNE, Paris (1991)
236 **Vergnaud, J** *Liquid Transport Processes in Polymeric Materials* Prentice Hall, New Jersey (1991)
237 **Walker, R, Knowles, M and Gilbert, J** 'Food additives and contaminants' *Fifth International Symposium on Migration* London 10–11 Sept 1987 Food Additives and Contaminants 5 (1988)

Glossary of packaging and ecological terms

Acryl(o)nitrile liquid with molecular structure CH_2=CH–CN. Monomer for the polyacrylonitrile (PAN) plastic resin. It is also an important compound in the synthesis of rubbery copolymers such as ABS rubber (acryl(o)nitrile-butadiene-styrene).

Additives as far as plastic packaging is concerned, a distinction is made between intentional and unintentional additives. Intentional additives include substances such as softeners, antioxidants, stabilizers, lubricants, UV-stabilizers, colourants, anti-block agents etc. added to the plastic mass before manufacture of the packaging material or container in order to improve the protective properties or to promote shaping etc. By unintentional additives we mean, on the one hand, substances from the reactive medium which remain in the plastic mass (reagents namely monomers, catalysts, solvents and their impurities), and on the other hand substances generated by heat treatment during shaping, by contact with atmospheric oxygen or the packaged products, and by the influence of light, and by the influence of any ionizing radiation as β-rays or γ-rays used for sterilization of the packaged content.

ADI abbreviation for 'Acceptable Daily Intake'. Maximum quantity of a substance which can be ingested daily without causing any secondary effects. Unit: mg substance per kg body weight.

Aerobic digestion aerobic bacteria require oxygen directly from the air: they cannot live and/or will not grow without oxygen. Oxygen is consumed by the bacteriological oxidative processes: water, carbon dioxide and heat are generated during aerobic digestion.

α-rays positively charged helium nuclei emitted by radioactive elements. Due to their charge and their high velocity, these particles (four mass units – two positive charges) can, in the event of impact, damage the cellular components and/or structure by ionization and/or by breaking down.

Anaerobic digestion anaerobic organisms will grow in atmospheres poor in oxygen and even in the absence of oxygen. During the oxidation processes for life, the necessary oxygen is taken from compounds rich in oxygen such as nitrates, nitrites and sulphates. Since there is a lack of oxygen, methane can be produced and the bacteriological process tends to be endothermic. Anaerobic digestion is characterized by a respiratory quotient – value higher than 1.

Antioxidant belongs to a group of products used for their protective action against oxidation. Antioxidants are added to plastics to reduce their degradation during the thermoforming process and exposure to weathering, the combined effect of light (mostly UV-light) and atmospheric oxygen. Most antioxidants used in the plastic sector have a phenolic structure.

Aseptic packaging a series of operations resulting in aseptic packaged goods (e.g. foodstuffs). Aseptic packages are produced by assembling under aseptic conditions both packaging wares and goods, each being sterilized separately. Aseptic packaging does not require a post-treatment to ensure preservation of the packaged food.

Bag (paper) a preformed flat or gusseted flexible container, longitudinally seamed and closed at one end, made from one or more plies of paper or from paper with other flexible materials or from laminated paper, coated or impregnated with some other material.

Bag-in-box is an inseparable combination of a box, mostly made from fibreboard, in which a flexible bag is attached to the inside. The inner bag is composed of minimum amounts of various materials to give protection and to provide the shelf-life each product requires. Bag-in-box systems are compact and normally provided with a handle; the weight being reduced, carry-home convenience has been enhanced. Most systems also have display facilities by the mounted spout. These systems are also used for aseptic packaging.

Becquerel (Bq) the number of radiations detected in an environment or emitted by a radioactive material per second. One Bq equals one disintegration per second; a radioactive source of 1 Curie (old unit) is characterized by an emission of $3.7.10^{10}$ Bq (see Curie) β (beta)-rays: accelerated electrons. By β-radiation we mean the emission of electrons due to the disintegration of a radioactive substance. By β-rays we mean a beam of electrons which is produced by an electric generator. β-rays can be classified following their energy-content: low energetic rays (3 MeV); rays with mid-range energy (between 5 and 7 MeV) and high energetic rays (10 MeV). β-rays can be used to treat foodstuffs for longer preservation (see also X-rays). The energy of the third group is limited to avoid induced radioactivity.

Biotope place where an animal or a plant is completely adapted to its environmental conditions; uniform habitat.

Blister packaging made of hard plastic. Generally the packaging consists of two items: the actual 'blister'-form is sealed onto a paperboard sheet, which contains the information. Blister packaging for small items becomes 'pilfer protected' by the increased size of the sheet.

BOD abbreviation for 'Biochemical Oxygen Demand'. The BOD (e.g. of waste water) indicates the quantity of oxygen consumed by biological processes in standard conditions (five days at 20°C) to break down (oxidize) organic 'biodegradable' products. BOD is expressed in mg/litre; it measures the degree of pollution by organic substances in aquatic systems.

Boltzmann's constant (cf. gas kinetic theory) is the factor k expressing the relationship existing between the kinetic energy of gas particles, their mass being m and their velocity v, and the temperature (expressed in degrees Kelvin)

$$\frac{1}{2} mv^2 = \frac{3}{2} kT$$

Numerical value: 1.38054 10^{-23} Joule/Kelvin.

Box (wooden) a lightweight, close-boarded or wood-based sheet material container, not normally fitted with girth battens.

CAP abbreviation for 'Controlled Atmosphere Packaging'. CAP finds application in the packaging of fresh food (fruits, vegetables etc.) which still consume oxygen during storage to produce carbon dioxide and water. CAP is performed by choosing the appropriate material and film thickness in order to keep the exchange of oxygen, carbon dioxide and to some extent water vapour between the inner and outer atmosphere under 'control'. As in the case of MAP, CAP occurs to extend preservation of food with the big difference that here, permeation of gases through the packaging wall during storage is fundamentally necessary.

Carton board paperboard with a grammage in the range of 250–900 g/m^2 used in making folding boxboard cartons.

Case (wooden) a heavyweight, close-boarded or wood-based sheet material container.

Cask a generic name for all-round, bilged, coopered wooden containers, constructed with staves, heads and hoops. A barrel is a cask of specific capacity.

Catalyst substance acting on the speed of a chemical reaction (mostly accelerating it) which can be recovered unchanged after the reaction. For polymerization reactions, special catalysts (peroxides) are used that become part of the polymer. The quantity of 'polymerization catalyst' used has an influence on the dimension and thus on the molecular mass of the final polymer.

Cellophane transparent packaging film resistant to gas transmission (in a dry atmosphere) made of regenerated cellulose. Cellophane is obtained from cellulose $(C_6H_{10}O_5)_n$ by the so-called viscose process: the cellulose insoluble in water is treated with some chemicals (NaOH, CS_2) to form soluble cellulose xanthate. By reaction with diluted sulphuric acid, cellulose is formed again (regenerated) in the desired form (films, fibres). Cellophane is nowadays coated with synthetic plastic resins in order to lower its permeability to water vapour and to enable heat sealing.

Cellulose colourless and tasteless high molecular natural product. General formula: $(C_6H_{10}O_5)_n$. Chemically cellulose is a polycondensate; from a large number of glucose molecules $C_6H_{12}O_6$ a natural high molecular product (MM: averages 106) is generated. Each bonded glucose has three free OH groups available for further chemical reactions. Cellulose is a vegetable material (main component of the cell walls). Wood contains approx. 40% cellulose and cotton approx. 95%. Cellulose fibres are the main component of paper.

Cellulose acetate compound derived from cellulose by reaction with anhydrous acetic acid (acetylated cellulose). In the cellulose, per unit of glucose, maximum three molecules of acetic acid can esterify with the existing free OH-groups to form the acetate. In practice, the mixtures used for plastic packagings contain on average 2.5 acetylated OH-groups. The acetate film is obtained by evaporation of the solvent, acetone, in which the concentration of the dissolved cellulose acetate reaches 60%.

Cellulose-(tri)nitrate compound derived from cellulose. General formula: $(C_6H_7O_5\text{-}(NO_2)_3)_n$. Often erroneously called (tri)nitro-cellulose. As for cellulose acetate, three free OH-groups from the glucose units of the cellulose can react with the nitric acid ($HO\text{-}NO_2$) to form nitrate bonds with water separation. The compounds slightly nitrated (10.5% N) are soluble in alcohol; the highly nitrated ones (12.5% N and 14.14% N) dissolve respectively in ether and acetone. In the packaging sector, cellulose nitrate is important as a coating material to improve the permeability properties of cellulosic films.

Child Resistant Closure see CRC.

Chromopaper paper mineral coated on one side with a binding agent providing a glossy surface. Chromopaper has excellent printing properties; it is used, for instance, for the packaging of chocolate bars.

COD abbreviation for 'Chemical Oxygen Demand'. COD (e.g. of waste water) denotes the amount of oxygen necessary to oxidize all existing organic compounds (biodegradable and others). Thus the COD of an industrial effluent is usually higher than its BOD. For domestic sewage the ratio COD to BOD amounts 1.5 to 2. COD is also expressed in mg/litre.

Compost is the final humus-like stabilized product obtained by biological processes from solid or liquid organic wastes present in household (or garden) refuse.

Consumer package a package containing a rather limited quantity (expressed in units, kg, litre) of goods to be delivered to the final consumer. Its design is determined by the needs of the distribution system and the normal consumption pattern whereby product quality can be insured. Synonyms: retail package – primary package.

Contamination limited environmental uncleanness characterized by the fact that natural biological cleaning process (decontamination) can occur. 'Contamination' does not imply a hazard to health, this in contradistinction with 'pollution' whereby the limits to natural decontamination processes are exceeded, when damages and health hazards occur or when ecological equilibria are drastically disturbed.

Copolymer polymer obtained by reacting of at least two different monomers together.

Corrugated fibreboard consists of one or more sheets of fluted paper (fluting) secured with an adhesive to one or more liners. The following combinations are usually fabricated:

- single-faced corrugated fibreboard: one sheet of fluting secured to a liner
- single-wall corrugated fibreboard (double-faced): one sheet of fluting secured between two liners
- double-wall corrugated fibreboard: two sheets of fluting separated by and secured to one flat sheet and faced on both sides by a liner
- triple-wall corrugated fibreboard: three sheets of fluted paper separated and secured to two flat sheets and faced on both sides by a liner

Crate (wooden) boarded case of open slatted construction.

CRC abbreviation for 'Child Resistant Closure'. Most so-called CRCs need two subsequent or combined operations to be unlocked. CRCs are labelled as such when following stipulations are fulfilled:

- 85% of a testgroup of children do not succeed in opening the closure without demonstration.
- 80% of the testgroup are not able to open the closure after a demonstration.
- 80% (especially for pharmaceuticals) of the testgroup may not achieve access to an amount of products which can cause serious injuries. Conversely, 90% of a testgroup of people

aged between 18–45 years must be able to open CRCs after reading instructions. CRCs are applied for packaging dangerous and/or toxic products as chemicals, detergents etc. used in household and pharmaceuticals.

Crystallinity percentage of crystalline structures in a material; e.g. the crystallinity of HDPE amounts to 80%.

Curie old unit (replaced by Becquerel Bq) indicating the number of disintegrations per second in a certain quantity of radioactive material. A one Curie radioactive source means that $3.7.10^{10}$ disintegrations occur per second in the selected matter. This is the radioactivity of 1 gram Radium or 0.88mg $Cobalt^{60}$ or 0.115mg $Caesium^{137}$.

Cushioning the use of a material or device to mitigate the effects of transport and handling shocks.

Cushion factor a measure to describe the efficiency of cushioning materials. The cushioning factor is the ratio of the maximum load per unit area to the energy absorbed per unit volume of the cushioning material. Because the ratio varies according to the thickness of the material, it is common practice to establish an optimum value of the cushion factor for a given material.

Duroplast see thermosetting resins.

EAN abbreviation for 'European Article Numbering'. 13 digit code for marketable products: two digits for the country of origin, five for the manufacturer, five for the product and one as a check digit. Country and producer codes are given by the EAN Authorities; the manufacturer chooses the product code. An eight digit alternative exists also. For automatic reading, these sequences of digits are converted according to the binary system into a bar code printed on the packaging (label).

Eco-balance of packaging synonym for life-cycle analysis of packaging. Eco-balances are produced by a step by step analysis of each substantial industrial process with regard to the environments being considered as a donor of resources and as an acceptor of wastes.

Ecological equilibrium is characterized by the harmonious coexistence of different groups of living organisms dependent on each other.

Ecology science studying the interrelationship of animals and plants with their environment; as far as the human being is concerned, ecology means also the study of the influences exerted on man by his environment.

ED_{50} abbreviation for 'Median Effective Dose'. Quantity of a substance provoking a visible effect on 50% of an animal test population.

Elasticity synonym for springiness. Ability of some materials to recover their original dimensions after loading or traction. Once past the elastic limit, the matter does not recover its original shape. Up to the elastic limit the length increase (elongation) is proportional to the applied load.

Elongation length increase occurring when a test strip of a particular material is submitted to traction. At first the elongation is proportional to the tensile stress but near the elastic limit the length increases faster than the applied load. Elongation stops when the strip breaks.

Entropy (cf. thermodynamics) ratio consisting in heat quantity supplied to a system during a very small reversible transformation over absolute temperature at which this heat exchange takes place. A reversible system (e.g. melting ice) reaches an equilibrium state when the variation of entropy tends to zero. In the case of spontaneous processes, an entropy increase means an increase in the disorder of the molecular structure. So, ice (stable state) melts when placed in hotter surroundings. In the liquid state the water molecules are free to move and the disorder is greater than in ice. Although the entropy concerns thermodynamic processes in physics and chemistry, the behaviour of living organisms and the consequences of heat exchanges in industrial processes can be explained by entropic considerations.

Environment all elements on earth having an influence on the existence of human beings and of a very large variety of animal and vegetable species.

Epoxy resins a group of plastic resins featuring a very good adherence to metal and a good resistance to chemicals. Hence they are used to protect tin plate. Epoxy resins are polycondensates from epichlohydrins (CH_2-OCH-CH_2Cl) and diphenols in the presence of NaOH. Generally epoxy resins are not toxic.

Ethylene an unsaturated hydrocarbon ($CH_2=CH_2$). Ethylene is the raw material (monomer) for the preparation of polyethylene (see PE).

EVA abbreviation for 'Ethylene-vinylacetate copolymer'.

EVOH abbreviation for 'ethylene-vinylalcohol copolymer'. It is generally obtained by (partial) hydrolysis of the acetyl-group present in ethylene-vinylacetate-copolymer. The 'ethylene' content and the 'vinyl' content in EVOH range from 29–48 mol% and from 71–52 mol% respectively. EVOH- and EVA-films have excellent gas barrier properties in dry conditions: therefore these films are laminated on both sides with polyolefins as water-resistant PP or PE. The presence of vinylalcohol groups (with formula: $CH_2=CHOH$) provides the copolymers improved properties concerning tensile strength and permeation rates in comparison to PE. EVOH is mostly used in coextruded laminates as an inner layer between nylon 6 and ionomer. EVOH is very stiff and transparent: at first, it looks like cellophane. EVOH is resistant to most solvents except alcohol. PVA (polyvinylalcohol) is soluble in water and finds special application with respect to this property (see also EVA).

Extrapolation inferring the validity of a law or the knowledge of a function beyond its actual limits. For instance, toxicity of products on human beings is determined by 'extrapolation' of data obtained by experimentation on animals.

Film a non-fibrous, non-metallic flat material, normally with a thickness lower than 250 μm.

Foil a rolled metallic material, normally less than 200 μm thick.

Foot English unit of length; 1 foot = 12 inches = $30.48.10^{-2}$ m.

Footcandle English unit of luminosity; 1 footcandle equals 1 Lumen per square foot (0.0929 m^2). 1 footcandle = 10.76 Lux (see Lux).

Frequency √ number of times that a periodic phenomenon occurs per unit of time (second): e.g. number of vibrations per second; the light frequency indicates the number of electromagnetic vibrations per second. Wave length multiplied by frequency of electromagnetic vibrations (light, radiowaves a.o.) is equal to light velocity (3.10^8 m/sec).

γ-ray one of the non-corpuscular radiations emitted by radioactive bodies. γ-rays are electromagnetic waves similar to Rîntgen rays but their wave-length is lower and their physiological impact (ionization and/or mutagenic effects) more severe. Most γ-radiations used for medicinal and for food preservation treatments are produced by ^{60}Co. This radiation consists of two energetic levels: 1.17 MeV and 1.83 MeV. The half-life time of ^{60}Co is 5.24 years, which means a decrease of 12% a year.

Glass a solid, transparent and breakable inorganic material. Chemical homogeneous solution obtained by fusion of various metal silicates in one or several vitroidal cooled non-metallic oxides. Due to its resistance against the attack of foodstuffs and of most chemicals, glass is a valuable packaging material. It is completely recyclable.

Glass temperature (Tg), also called 'glass transition temperature'. Parameter specific to linear polymers. Tg indicates the temperature at which a structural modification occurs in solid linear plastic resins: when the temperature rises above the Tg, the glassy amorphous structure becomes rubbery. Polymers can be characterized by their glass transition temperature, although this is dependent upon the presence of additives. Tg for PVC is 87°C: appropriate plasticizers will reduce Tg to 40°C, as in the case of PVC-shrinkfilm.

Glassy state a (for glasses and linear polymers) specific amorphous, non-crystalline state: glass can be considered as a supercooled liquid whereby the molecules remain in disorder. Synonym: vitreous state. In the case of thermoplastics, the glassy state occurs at that temperature where the linear molecules are no longer susceptible to rotation.

Gy abbreviation for 'Gray': unit for the absorbed amount of energy from radioactive radiations (see RAD). 1 Gy = 100 RAD; 1 Gy applies to a dose of 1 Joule per Kg of substance.

Heat-sealing joining by fusion under heat and pressure. Heat-sealing is widely used for making sachets and pouches and their closures.

Heterogeneous consisting of dissimilar constituents.

Horsepower (HP), Power to displace 75 kg through 1 metre in 1 second (1 HP = 736 watts)

Household refuse is all solid waste issuing from normal households. As in towns, wastes from restaurants, shops, markets, small companies and administrative activities (mainly paper) are also often supplemented by house-to-house collections. Its composition is not only very heterogeneous but it may also contain toxic substances. Actual packages may represent about 30% by weight and 50% by volume of household wastes. Weight and volume reduction are planned by stimulating the reuse and/or recovery of materials.

Ionizing radiations corpuscular or electromagnetic radiations, the energy content of which is high enough to provoke the ionization of materials. This causes one or several electrons to leave the atoms; in the case of organic molecules (including those from living organisms), ionization is generally complex, as either break-down reactions occur simultaneously, or new structures appear by recombination. Ionizing radiations have a germ-killing action and can thus be used to sterilize foodstuffs and their packagings.

Ionomers plastic resins, the linear molecules of which feature, at regular intervals, chemically reactive branches where the polymer chains are strongly bound to each other by means of metallic ions (ionic bonds are stronger than Van der Waals' attraction forces between identical molecules, called cohesion forces). The polymerized chains are thus linked to each other laterally. Unlike duroplasts (three-dimensional structure due to covalent bonds), ionomers are flexible, malleable when heated, although resistant to traction.

Joule unit of work (energy); dimensions $kg.m^2/sec^2$.

Kraft paper light brown-coloured paper featuring a high resistance to traction, obtained from unbleached sulphate pulp. Grammage between 50–250 g/m^2.

Laminate the product resulting from combining the complete surfaces of two or more webs. Also, any combination of coextruded films or of films, foils and papers are considered as laminates.

LCA abbreviation for 'Life-Cycle Analysis'. During the last decade LCAs of packagings have been made mainly to investigate the material and energy needs on the one hand and the possible environmental impacts during the whole life-cycle on the other hand. The aim of the LCA of packaging from 'cradle to grave' is to discover the relative importance of all production processes packagings are involved with, and their environmental impact. Also, LCAs aim to reduce material and energy losses and to improve the production processes.

LD abbreviation for 'Lethal Dose'.

LD_{50} abbreviation for 'Median Lethal Dose'. Dose of a substance the ingestion of which causes 50% death among an animal test population. LD and LD_{50} are expressed in mg of substance/kg of body weight.

Life-cycle analysis of packaging see LCA.

Light absorption decrease of the light intensity in passing through a material (packaging). Absorption is the logarithm of the ratio I_0/I where I_0 and I are respectively the light intensity before and after passing through a medium. Absorption depends upon the material and the wave-length of the light.

Light flux quantity of light emitted or received per second. Unit: Lumen (Lm). The light flux emitted by lighting lamps depends on the power (in Watt) and the lighting system: incandescence lamps produce 10–15 Lm per Watt, fluorescence lamps 40–80 Lm per Watt.

Lignin a strong, rigid, complex and not easily biodegradable natural substance coating the cellulosic cell walls of some plants, enabling transport of the ascending sap. Wood contains on average 30% lignin. Typically newsprint made from mechanical pulp will become yellow and friable under the influence of solar light and oxygen because of the presence of lignin.

Liner the loose inner lining of a container. Liners can be flat or pre-shaped e.g. for arranging fruit in order.

Litter refers mainly to packaging waste. Defined in the literature as waste dumped or lying on an inappropriate ground.

Lux unit of luminosity (light-intensity); 1 Lux equals 1 Lumen illumination on 1 square metre area.

MAP abbreviation for 'Modified Atmosphere Packaging'. It is a technique to extend preservation of packaged fresh and precooked food by replacing the principal alteration agent oxygen inside the package by the gases nitrogen and carbon dioxide. Exceptionally, small amounts of oxygen must remain present (colouring principle of meat) and concentrations of carbon dioxide up to 25% will inhibit the growth of micro-organisms. MAP being not a sterilization technique, the initial microbiological quality of the food and the low permeation to gases of the packaging material remain the most important parameters for extended preservation. Laminates of PE and PA or PE and PET are commonly used for MAP.

Marking of packages application of numbers, letters, labels, tags, symbols or colours for identification and handling instruction during transport, storage and use.

Mechanical wood pulp aqueous suspension of small wood fibres obtained from debarked soft wood logs when reduced to small chips at high pressure and temperature in the 'wood grinding machine'.

MeV 1 MeV= 10^6 eV, unit to describe the energy of radiations (α-, β-, γ- and X-rays). 1 eV= $1.6.10^{-19}$ Joule

mg milligram = 10^{-3} gram.

Microgram (μg) 10^{-6} gram.

Microwavable packaging is obtained by using materials resistant to high temperature and the influence of the electro-magnetic waves (wave-length approx. 28 cm) produced in the microwave-oven. Microwavable packaging is designed to contain the food during their cooking and/or heating in the oven. Aluminium foil when thin and uniformly distributed may be applied especially when dual-oven packagings are planned (to be used in the conventional and microwave-oven.)

Migration transfer of substances (migrants) from the packaging surface into the packaged product (in the case of foodstuffs). A distinction is made between 'global' migration – sum of all migrants (known + unknown) – and 'specific' migration – meaning the transfer of particular constituents, namely metallic ions, monomers, plasticizers etc. from the packaging into the packaged foodstuff. Migration of constituents can:

(a) occur spontaneously (migration of monomers and/or volatile products from the packagings) or
(b) be induced by contact with the packaged content (mostly liquid) or
(c) be non-existent; this occurs only in some particular cases (appropriate combination between material and packaged product).

The legislators have put a limitation on global migration for most types of plastic packaging. It must not exceed 10 mg/dm^2 (60 mg/kg). Limitation also exists for specific additives (see positive list).

Mil English unit of length used to indicate the thickness of foils and films; 1 mil = 1/1000 inch = $2.54.10^{-3}$ cm = $25.4.10^{-6}$ m.

MM abbreviation for 'Molecular Mass'. Number expressing the ratio between the mass of one molecule of the compound in question and 1/12 of the mass of a carbon atom.

Molecule smallest particle of a compound or of an element, featuring all physical and chemical properties of the substance in mass, electrically neutral.

Monomer general term assigned to a range of low molecular compounds linked by polymerization, by polyaddition or by polycondensation to form bigger molecules called polymers; monomers are thus the specific building bricks of polymers (and copolymers).

Multi package several primary packages collated into one retail unit.

Multi-trip package a package designed to make several times the packer-consumer cycle.

Mylar trade name of a polyester produced by condensation of glycol and terephthalic acid.

Newton unit of force. Dimensions: $kg.m.sec^{-2}$.

Nm abbreviation for nanometer. Unit of length; $1 \text{ nm} = 10^{-9}\text{m}$.

Nylon assigned to plastic resins also called polyamides. Originally Nylon was a trade name (du Pont de Nemours) for the polyamide produced from hexamethylenediamine and adipic acid (Nylon 66).

Nylon 6 polymer produced by polycondensation of an aminocarbonic acid, (aminocaproic acid), or by heating its derivative caprolactam.

Nylon 66 polymer prepared by condensation of a dicarbonic acid (adipic acid) and a diamine (hexamethylenediamine). Index 66 (read six,six) indicates the number of carbon atoms respectively in the acid and the diamine.

OVAM abbreviation for 'Openbare Afvalstoffenmaatschappij voor het Vlaamse Gewest' (Public Organization for Waste Management of the Flemish Region). Organization created by the decree dated 2 July 1981, competent in many matters relative not only to waste disposal but also empowered to enforce regulations aiming to avoid the production of some particular wastes.

Oxidation reaction with oxygen.

PA abbreviation for 'polyamide'. Polyamides have an amide structure (-CO-NH-) in their main chain. Nylons are typical representatives of the PAs. They can be manufactured by polycondensation of aliphatic diamines with dicarboxylic acids. Nylons are marked by the number of carbon atoms in both the diamines and dicarboxylic acids. Nylons 66 relates to hexamethylenediamine and adipic acid, both having six carbon atoms in their molecules. Polycondensates of hexamethylenediamine with sebacic acid (10 carbon atoms) or with dodecanedioic acid (12 carbon atoms) produces nylon- 610 and nylon 612 respectively. When aromatic diamines as metaxylylene diamine (MXD) or aromatic diacid as isophthalic acid (IPA or I) are used, the polymer is called by the abbreviation followed by the carbon number of the partner:

Nylon-MXD6 stands for metaxylylenediamine and adipic acid.Nylons can also be produced by polycondensation reaction of molecules possessing both an amino and an acid group or their cyclic derivatives (lactams). Nylon 6 stands for polycaprolactam; Nylon-11 and Nylon-12 stand respectively for poly-omega-amino-undecanoic acid, and the polymer of the lactam12aminododecanoic acid. PAs have relatively high melting points (250°C), very good gas barrier properties and good resistance to most liquids.

Package the result of a complete series of packaging operations or a unit consisting of a number of such products i.e. the package includes the product packed. Packages differ from other wrappings by the presence of some sealing device, to be broken for access to the content (insurance for producer and consumer). Synonyms: pack, packet.

Packaging the whole series of operations involved in the preparation and wrapping of goods for handling, distribution, transport, storage, delivery and sale to the final user. The term includes preservation, packing, marking and unitization of the goods.

Packaging accessory supplementary components for packaging (e.g. cork, label, rope, insert).

Packaging material substance used to manufacture packaging wares. Examples: glass, tinplate, PE plastic resin, paperboard, etc.

Packaging ware main component of the package, container, box, bottle, jar, etc. Synonym - Packaging(s)

Packaging waste general term to define a series of solids resulting from used packaging (primary, secondary or tertiary) or materials of packages which have completed their main function of ensuring sound delivery of a product and which are to be discarded. A distinction is made between:

- Industrial scrap: unused packaging materials and/or wares rejected by the manufacturer for breakage, damage or bad processing and which are directly recycled.
- Post-operation materials: unused scraps, rejected packages or parts of them, generated during their manufacture, which are not recycled in the plant but directly collected for that purpose elsewhere.
- Post-consumer waste: used packaging materials as left by the consumer. It is mostly spoiled, unclean, very heterogeneous and widely scattered.

Packing the operation of packaging by which the product is enveloped in wrapping and/or enclosed in containers or otherwise secured.

Pallet load board used to carry one or a number of combined packaged items as a unit load for handling, distribution and storage. A pallet comprises a flat upper deck with a lower deck designed to permit movement by materials handling equipment e.g. fork trucks.

PAN abbreviation for Polyacrylonitrileplastic produced by polymerization of acrylonitrile.

Paper (see also paperboard) sheet material made from fibrous stock (mainly wood fibre – also cotton), fillers (mainly sulphates, clay, titaniumdioxide) and binding agents (e.g. phenolic resins). Generally so-called paper has a grammage below 225 g/m^2.

Paperboard general term for all sheet material made from fibrous stock (mainly wood fibre) with a grammage of 225 g/m^2 or greater. Material below 225 g/m^2 is generally paper.

Parchment paper paper treated briefly with a 40% solution of sulphuric acid. As a result, the cellulose molecules are partly decomposed and absorb some water which modifies the fibrous structure. The paper becomes translucent.

PC in the packaging sector abbreviation for 'polycarbonate'. PC-polymers are amorphous, provide excellent impact strength, heat resistance (melting point up to 230°C) and transparency (90% for visible light; most UV-light absorbed). The brittle temperature amounts to –100°C: PC has a significant good resistance to heat and cold and is hence used for the production of dual ovenproof trays and of containers for hot fillings to be kept deepfrozen. Polycarbonates are also used in laser-optical data-storage systems (e.g. compact discs).

PE abbreviation for Polyethylene. Polyethylene is a plastic resin produced by polymerization of the ethylene monomer (CH_2=CH_2). According to the preparation process used, either low density LDPE or high density HDPE plastic resins are obtained. LDPE production results from a high pressure process, and HDPE from a low pressure process; LD and HD refer now only to the density and no longer to the process. Density ranges between 0.910–0.925 kg/dm^3 for LDPE and between 0.941–0.965 kg/dm^3 for HDPE. Generally PE plastic resins burn without producing soot; but have a strong tendency to form burning droplets.

Period interval of time between two successive identical states of a periodic phenomenon.

Permeability permeation by gases or vapours.

Permeation rate the quantity of a gas or vapour diffusing per unit of surface (m^2), unit of pressure (bar) and unit of time (day) through a sheet the thickness of which is 100μm. Dimension: g.100μm/m^2.day.bar.

PET abbreviation for polyethyleneterephthalate. PET belongs to the group of polyesters, being produced by polycondensation of para-terephthalic acid (formula: HOOC-C_6H_4-COOH) and glycol (di-hydroxy ethane with formula HO-CH_2-CH_2-OH). Hence, the chemical structure of PET can be visualised as -(OCH_2-CH_2-OOC-C_6H_4-CO)$_n$-. PET molecules can easily be oriented by the blowing process, so that PET-wares (e.g. bottles) show a very low permeation rate to gases, which make them suitable for packing carbonated drinks.

pH value symbol indicating the acidity of an aqueous solution. Measure of the concentration of hydrogen ions (CH+) in aqueous solutions. pH = –log CH+ where CH+ = number of equivalents per litre. pH of acid solutions is below seven, pH of alkaline solutions is above seven. Pure water has a pH = six to seven (neutral).

Planck's constant (cf. radiation energy) is the factor h expressing the relationship between the energy E of electromagnetic waves and their frequency: E = hv. Numerical value: $6.6262.10^{-34}$ Joule/second.

Plasticizers range of substances added to plastic resins in order to increase their malleability and thus to ease the processing of the final products. Plasticizers are mainly liquids of high boiling point and low vapour pressure. Important examples are the esters of phthalic acid (phthalates).

Plastics group of organic, high molecular products obtained either derived from natural high molecular products or in a fully synthetic manner from smaller molecules.

Plexiglass plastic from the acrylate group obtained by polymerization from CH_2=C-$COOCH_3$ (methylester of methylacrylacid or methylmethacrylate). Hence the abbreviation PMMA – polymethylmethcrylate – is used for this polymer.

Pliofilm almost colourless transparent film (or foil) produced from a hydrochlorinated natural rubber. The chloride content ranges from 28 to 31%; the film is obtained either by calendering of the rubber mass or by the evaporation of the solvent toluene from a solution of rubber. Pliofilm is suitable for heat-sealing (thermoplastic). It is hardly flammable and resists fat, oils and water. Hence pliofilm was used for packaging purposes.

PLU abbreviation for 'Product Look Up' normally followed by a number. PLU is thus a code mentioned on the packaging (label). The cashier typed this number to call back the price from the computer memory and issue the bill.

Poisson coefficient indicates the stricture of a test strip made of some material as a function of its elongation during a tensile test.

Pollution we mean here severe contamination (e.g. of an aquatic system with a biological equilibrium of species) insofar as the natural decontamination processes began to operate out of acceptable limits of variations in that biosphere. When pollution occurs, the biological equilibration processes are deeply disturbed: the normally present great variety of animal species (sometimes also plant species) occurring at a low density of population are replaced by a limited number of species with high densities of population. In the most polluted systems, generally bacteria survive best. Besides their impact on the natural environment, pollutants can also damage buildings and works of art.

Polyamide see PA and Nylon.

Polyesters generic name of a group of plastics prepared by condensation of polyalcohols with polycarbonic acids. Polyesters usually burn with a sooty odour.

Pouch a closed rectangular bag, such as used for potato chips.

PP abbreviation for 'Polypropylene'. Plastic prepared by polymerization of propylene (CH_2=CH-CH_3). Like PE, PP burns without producing soot, but with a lesser tendency to form burning droplets.

ppb abbreviation for 'parts per billion'(µg/kg, µl/m^3).

ppm abbreviation for 'parts per million'; mg/kg, and ml/m^3 are equivalent.

PS abbreviation for 'polystyrene. Plastic prepared by polymerization of styrene (CH_2=CH-C_6H_5). As the benzene ring is rich in carbon, PS produces a lot of soot when burning.

PSI abbreviation for 'pounds per square inch'. English unit used to express the applied tensile stress.

PTFE abbreviation for 'polytetrafluorethylene'. See Teflon.

Pulping (cf. paper-making) process liberating the cellulose fibres from wood.

- Chemical pulping: process where wood chips are submitted to the action of chemicals in boilers at high temperature in order to make lignin soluble.
- Mechanical pulping: process where debarked logs of wood, sawn to the desired length, are pressed to a grindstone or wood chips are passed between rotating discs with plenty of water to obtain a suspension of wood fibres.

PVC abbreviation for 'polyvinyl chloride'. Polymer belonging to the vinyl resins group, produced by polymerization of the vinylchloride monomer (VC: $CH_2=CHCl$). Unlike many plastic resins of the packaging sector, PVC submitted to heat treatment (pyrolysis) is not decomposed into its monomer but into a complex range of breakdown products that include hydrochloric acid (HCl).

PVdC abbreviation for 'polyvinylidene chloride'. Polymer produced by polymerization of the vinylidene chloride monomer (VdC: $CH_2=CCl_2$); General formula: $-(CH_2-CCl_2)_n-$. The polymer chain is made up of VdC-units added head to tail and the crystal structure of the polymer is fairly well established. The importance of VdC as a monomer results in its ability to copolymerize with other vinylmonomers (see Saran). The most valuable property of PVdC and its copolymers is low permeability to a wide range of gases and vapours – PVdC and copolymers find wide application as coating materials.

Quality concept expressing to what 'degree' the properties of a product meet some requirements fixed in advance. Usually the quality of packagings can be illustrated by several physical parameters. In the general case of foodstuffs, and in the particular case of packaged foodstuffs, a wide range of non-quantitative factors (emotional) are taken into account in the overall appreciation of the quality.

Rad abbreviation for 'Radiation Absorbed Dose'. Energy of rays absorbed by an irradiated material. This unit applies to a dose of 1/100 Joule per kg of material.

Recycling of packaging materials Treatments aimed at reintroducing into the manufacturing process, used packaging materials in order to produce new packaging wares or products. In this case, the manufacturing process is completely performed again. When recycling glass packagings, the broken glass ('cullet') normally separated by colour is dumped again in the smelting furnace. Metals can be recycled without great difficulty. Generally recycling leads to savings in material and energy; recycling of plastics (thermoplasts) is theoretically feasible but practically there are several problems namely: the occurrence of structural modifications (the structure of the collected plastic differs from the original one due to thermodegradation and weathering), the wide variety of materials (whether or not combined are difficult to separate), the presence of impurities (spoiled wastes are generally difficult to purify). Plastic scrap originated from the thermoforming process is generally recycled at a rate of 20%. Post-consumer plastic wastes may not be recycled for food packaging purposes. Paperboard can be recycled to some extent (see paper pulping). Usually, packaging paper will not be recovered for recycling purposes.

Rem abbreviation for 'Röntgen Equivalent Man'. This unit represents the absorbed dose of one Röntgen when exposed to radioactive radiation. The maximum allowable dose within a human lifetime is estimated at 250 Rem (five to maximum 12 Rem per year).

Resonant frequency phenomenon occurring when a body vibrates under the action of external forces at a frequency equal to its natural frequency.

Retort pouch A heat-sterilizable package produced from a two, three or four ply flexible laminate which when sealed is as hermetic as a can or jar; and is sterilized by retorting.

Röntgen measure of the dose and intensity of the radioactive radiation received by a body per unit of surface. An individual located at 0.91 m from a source emitting one Curie receives per hour a dose equal to one Röntgen.

Röntgen-rays see X-rays.

RQ abbreviation for 'Respiratory Quotient'. Ratio between the quantity of carbon dioxide produced (CO_2) and the quantity of oxygen (O_2) consumed during biological oxidation processes. The RQ quotient is an important parameter for composting processes: if greater than one, it indicates anaerobic processes, if smaller than one, there are elements other than carbon combining with oxygen (unwanted oxidations).

Sachet a small package, originally cushion-shaped, made from flexible film. The term sachet now also refers to other shapes.

Sack (paper) a container made essentially from one or more flattened tubular plies of paper closed at least at one end, possibly in combination with other flexible materials to provide the properties required for filling and the goods distribution chain.

Saran trade name for a copolymer of vinylidene chloride (VdC 85%), vinyl chloride (VC 13%) and some acrylonitrile (AN 2%).

Sheet plastic a non-fibrous, non-metallic flat material, the thickness of which is normally equal to or higher than 250 µm.

Shelf-life the period during which a package and its content do not deteriorate beyond acceptable limits when stored and displayed as designed and prescribed. Synonym: package life.

Shrink film is prepared by stretching a film when heated above its glass transition temperature and subsequently cooled while stretched. Shrink films have the property to shrink back to their original shape by reheating above their Tg. Hence Tg of shrinkable materials must be higher than room temperature. Shrink films are widely used to produce tightly wrapped packs.

Skin pack a pack that encloses an article on a suitable support with a film that is vacuum-formed to the contours of the packed article.

Stabilizers substances present mainly in PVC to create a chemical bond with the hydrochloric acid generated by decomposition during processing and use; hereby, the plastic material is protected against further attack.

Sterilization treatment aiming to make goods free of microbes.

Stress cracking cracks occurring in moulded plastic wares when submitted to high loads or to vapours by which the polymer tends to swell. Stress cracking occurs preferentially on places where residual forces remain 'frozen' in the objects (e.g. in necks and bottom of bottles). Leakage can occur through these cracks. White spirit is a stress cracking inducer for PS; xylene and toluene for PE.

Stretch film films made from PE, EVA and PVC (with Tg below room temperature). Like shrink film, stretch films are competing in the fields of collating, pallet overwrapping and retail wrapping. The general principle of stretch wrapping is that the film is stretched around the object(s) to be wrapped and then sealed by pressure.

Styrene colourless liquid, monomer for the production of polystyrene. Chemical formula: CH_2=CH-C_6H_5.

Sulphate pulp cellulose fibres obtained from vegetable materials (mostly wood) by the action of hot alkalis ($Ca(OH)_2$, NaOH, etc.). The chemical treatment, called chemical pulping, dissolves the non-cellulosic components (lignin etc.). Sulphate pulp is light brown-coloured but may be bleached.

Sulphite pulp cellulose fibres obtained from wood heated and treated with calcium hydrosulphite ($Ca(HSO_3)_2$). The non-cellulosic components (lignin etc.) are dissolved and the cellulose remains almost unchanged. Sulphite pulp is slightly coloured but, unlike sulphate cellulose, is easily bleached.

Surlyn A first commercial ionomer, similar to PE. The printability of Surlyn is easier than PE by the presence of ionic groups (see also ionomers).

Tack The ability of an adhesive to form a bond of reasonable strength immediately after the surfaces to be adhered are brought together and generally before the adhesive has set.

Tamper evidence packages are normally provided with a closure combined seal, to be broken for content's access. In most cases this seal and/or closure cannot be restored after breakage: tamper evident packagings give insurance for both the manufacturer and the consumer.

Tamper evident a TE container or package has an indication or barrier to entry which if breached or missing can reasonably be expected to provide visible evidence that it has been tampered with.

Teflon trade name for the fluorinated polymer (polytetrafluor ethylene PTFE) very resistant to acids, not very permeable to gases and water vapour. Chemical formula: $-(CF_2-CF_2)_n-$.

Tensile strength maximum allowable stress applicable in standard conditions to a test strip of a material (packaging material) before rupture.

Thermoplasts high molecular products with a linear structure becoming plastic when heated (120–150°C) enabling them to be given the various shapes.

Thermosetting resins (duroplasts) three-dimensional high molecular products which, unlike thermoplasts, do not melt when heated but are broken down (> 350°C).

Throw-away packages packages designed for a single use. Synonym: single-trip or one-trip.

Tin plate rolled steel plate (thickness approx. 300 µm) coated with tin. The thin protective layer was originally obtained by immersion in molten tin now by electrolytic processing (electrolysis of an aqueous solution of tin salts). The tin layer is very thin (thickness approx. 10 µm which represents 1.5–1.7% of the total weight) and can be removed for recycling. The remaining iron can also be recycled. Tinplate, coated on one or both sides with a synthetic resin, is used to manufacture cans and other metal packagings.

TPX TPX belongs to the same family as polyethylene and polypropylene: it is also a polyolefin and thus resistant to acids, alkalis and many solvents. TPX is softened by hydrocarbons and differs from PE and PP by its clarity, its density 0.83 (lower than LDPE) and its permeability for gases and water vapour which is higher than that of either PE and PP.

Transition temperature (see glass-transition temperature).

Transport package package specially designed for transport and/or for a transport mode.

UHT abbreviation for 'Ultra High Temperature'. Process used to sterilize milk. By short (two to four seconds.) heating to a very high temperature (130–150°C), followed by rapid cooling, milk becomes sterile.

Unit package does not contain smaller packaged units; dependent on its size, a 'unit package' can be a part of a 'collective package'. Every unit package can be considered as a consumer package, but the opposite is not always true.

UPC abbreviation for 'Uniform Product Code'. Code number for products on the American market. See EAN.

UV-light abbreviation for 'Ultra Violet Light'. Region of the electromagnetic spectrum. UV-light covers practically the wave-lengths ranging from 150 to 290 nm.

Vapour pressure pressure exerted in a closed system by the gaseous phase of a substance in equilibrium with its liquid form.

Vinyl chloride gaseous compound (VC: $CH_2=CHCl$) important as monomer for the production of PVC.

Vinyl resins generic name for plastic resins whose monomer contains the vinyl radical $CH_2=CH$. The compounds with the general formula $CH_2=CHX$ (X being something else than hydrogen) derive mainly from acetylene ($CH \equiv CH$). PVC, PVdC, PS, PAN are typical examples of vinyl resins.

Watt unit of power: work done per unit of time. Dimension: Joule per second.

Watt-hour unit of energy; 1 Wh equals 3600 Joule.

Wave-length (v) distance between the peaks or troughs of two successive waves or the distance that a wave motion covers within one period. Unit of wave-length for light is either the nanometer (nm = 10^{-9}m) or the Ångström (Å = 10^{-10}m).

Wrapping the process of wholly or partly covering an article or of a group of articles with a flexible material (film, paper or aluminium foil). Wrapping also differs from packaging by a lack of functions which are specific to packaging.

WVTR abbreviation for 'Water Vapour Transmission Rate'. It means the rate at which water vapour passes through a physical barrier. WVTR is expressed in grams per square metre per 24 hours under specified conditions of temperature and relative humidity.

X-rays synonym for Röntgen-rays. X-rays are electromagnetic waves produced by bombarding metals with energy-rich electrons. The energy of the X-rays depends upon the kind of metallic target and the energy of the electron-beam: their range is limited to 5 MeV. Most have a wave-length ranging from 10^{-8} to 10^{-10} m.

Young's modulus quantifies the relationship between the applied tensile stress and the relevant elongation. Young's modulus E depends very much on the temperature and it is a major datum to determine whether plastic resins are suitable for packagings.